LONG JOURNEY

Edward Hill

In the year 2065, we are passengers in a space cruiser, on the way to Planet Alpha-1, four light-years away. Our mission is to do some exploring to see if there are any new sources for minerals or other items that will be useful on Planet Earth. We will be traveling at four times the speed of light, and it will take close to a year for the trip out. And another year for the return, the ship is loaded with engineers, physicists, and academic types, so we have plenty of talent to do what we need when we arrive.

We invented a new game. 3-D Basketball. During times of zero-G we can float through the air. 3-D Basketball is fun to play and fun to watch.

But completing the trip is no longer a given. We have Problems.

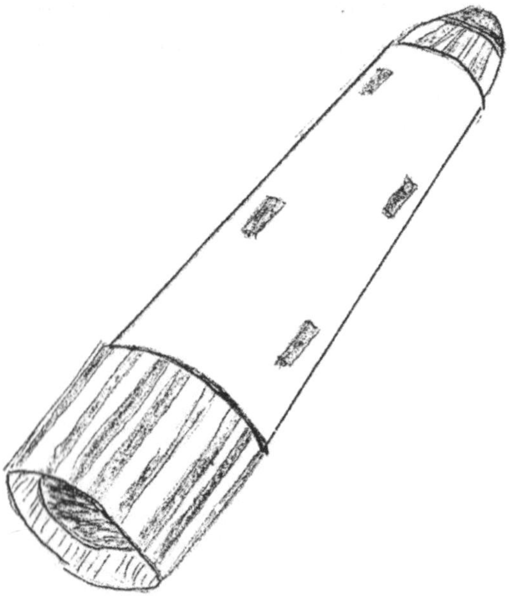

Table of Contents

Chapter 1 – The Beginning

Background. This is the year 2065. Increasing populations and diminishing resources on Planet Earth caused the International Syndicate, an important part of the United Nations, to start looking for new real estate. Explorations to other planets in our galaxy had resulted in some benefits, but not nearly enough. Our cruiser will be making the very first trip outside our galaxy, and the first distant voyage at light-plus speed. The travel time was to be a little more than ten months each way. With two or three months for exploration, and the total mission was planned for two years. The ship carries two hundred passengers, primarily physicists, scientists, and academicians, carefully selected in order to evaluate new finds. The ship's operations and maintenance crews are another twenty-five.

The farthest trip so far had been to Pluto, about eight billion miles from Earth, just four years earlier. Our new destination, Planet Alpha-1, is a lot farther; four light-years away, or twenty-three TRILLION miles. In order to meet our planned schedule of two years, most of our trip will need to be at light-8, or eight times the speed of light (L8 for short).

Loading. The ship's crew, highly qualified employees of the National Space Exploration Administration (NSEA), have been on board SC12 for a week, preparing for lift-off which is scheduled in another two days. The name of our ship is SC12, short for Space Cruiser Twelve, a two-year old modern-technology space craft.

SC12 is currently docked at Columbia, which is orbiting twelve thousand miles above Earth. Our very special passengers are gathered at the NSEA campus, at the ocean front in North Carolina, and by now have listened to several more briefings on the mission, as well as instructions for using the SC12 equipment, with which they will soon become very familiar. Each passenger has also been subjected to what is undoubtedly the most complete physical they have ever received. After checking in, they will be boarding the shuttle for the three-hour trip to rendezvous with and transfer to Columbia and then to board SC12. Loading is always the first step and usually takes several days. We have one shuttle, and it is limited to one hundred and fifty passengers at a time, so two trips were scheduled for this purpose.

Today's passengers are divided into small groups for checking in and then they will be escorted to the shuttle. The first crew member they meet is the SC12 Officer of the Watch, Lieutenant Bob Ventner, who is standing by to receive the passengers.

Bob stepped forward and started his routine, "Welcome to NSEA's shuttle craft. I will need some information before you can board."

"Hurry up and wait, huh."

"Not really. This will only take a couple of minutes. You're first. Name?"

"Mary Smith."

"Age?"

"Thirty-Two."

" Where are you from?"

"San Diego…California."

"All I need is the country. U.S."

"Occupation?"

"What do you mean? I work for NSEA, like most of the people here."

"No. What is your profession? Or your special skill that brought you here?"

"Oh, I see. I'm a scientist. Actually, an aeronautical engineer."

"Got it. Now wait here for your escort. He'll be here in a few."

Motioning for the next person, "You're next. Name?"

"Raymond Hilt."

"Age?"

"Twenty-nine."

"Country?"

"Keokuk, Iowa, USA."

"Really? Howard Hughes was from Keokuk, right?"

"Not really. His grandparents settled there in their later years. Howard visited them a number of times when he was a teenager, and a few more times when he was a young man."

"Neat. Occupation?"

"Howard was a lot of things. Aviator…"

"Not Howard. You. What is your occupation?"

"I'm a former Navy jet pilot. Just an average aeronautical engineer now."

"Thanks. (And a wise guy, to boot!) By the way, nobody on this flight is just average."

Bob motioned for the wise guy to move on. "Next."

"Robert Jessen, but most people call me Red. I grew up in Hamburg, Germany, moved to San Diego twenty years ago. Forty-eight years old, and currently I'm an electronic engineer."

"And what were you before?"

"Teaching physics. At University of Hamburg and then Oxford University."

"Wow. That's impressive. Why are you here?"

"Teaching physics is OK, but a bit boring. My wife left me a couple of years ago, and I started looking for a change. Exploring distant planets should be a little more exciting."

"You are probably right there. Welcome aboard."

Bob finished his notes, closed his clip board, and addressed the group, "Thanks guys. That's all I need. This is Lieutenant Byron Williams. He will escort you to your pre-assigned seats."

"Hi gang. My name is Byron. Stay close now, and I'll give you a little tour on the way. Has anyone been on a shuttle flight before?" Several heads wagged left and right. "Nobody. OK. The actual transit time for our twelve thousand and four hundred miles is less than two hours. But that includes fifteen minutes to accelerate from lift-off to cruising speed, and then a half-hour for deceleration and docking maneuvers. After you have found your seat, we want you to sit down and get familiar with the lock and unlock positions of your seat belts. Then you will have some time for a trip to the head. Any questions so far?"

Several hands came up. Byron pointed to one of them who said, "What is the head?"

Byron smiled, "Oh yes, "head" is the military equivalent for restroom. Many of us are former military, and we have continued to use that term." Byron looked at his notes, then continued, "Just like on an airliner you will be told to fasten your seatbelt before the lift off. You will be permitted to unlock your seatbelt for about an hour, and then you will have to be locked in again for the last hour. We have two shuttle flights to Columbia. One is your flight this afternoon, and the other is scheduled for tomorrow morning."

Receiving and boarding passengers on the shuttle continued for nearly another hour. New arrivals concentrated on stowing their carry-ons, visiting with old friends, and getting acquainted with some new friends.

The Shuttle Flight. The shuttle looked a lot different from the several space cruisers that were parked in the NSEA Campus. The cruisers were parked standing on their tails, while the shuttle was lying flat on the ground. Byron was talkative, pointing out items of interest. He could not help noticing the passengers were becoming more nervous the closer they came to the short stairway leading to the large opening in the side of the ship, not unlike the stairs for boarding an airliner. Byron was doing all the talking, while most of the others were silent. This changed, however, when they stepped inside and saw the layout was, again, similar to that of an airliner, and then conversations seemed to continue from where they left off.

Most of us are carrying briefcases or overnight bags. None of these are allowed in the seating area. Byron showed us where to deposit these and told us we could pick them up again when we are ready to transfer to SC12. Twenty minutes later, all one-hundred passengers were on board, and a polite buzzer sounded. At the same time, we saw the signs turn on, "Seat Belts," and there was an announcement, "Welcome aboard everyone. This is your Shuttle Captain speaking, and it is time for everyone to fasten their seat belts. We cannot lift off until all passengers are belted in. Right now, four, no, three belts are not fastened."

A minute later, the lights told us all seat belts were fastened. Then we heard Byron's voice again, "Stand by for lift-off."

Just a few minutes earlier, we heard a low growling sound from warm-up, and now that sound became higher pitched - and louder. Many of the passengers placed their hands over their ears, but a few seconds later the pitch was high enough to be out of range of our hearing. We could feel the lift-off, and then we felt the shuttle change direction, and then there was another message from Byron, "Everyone, please make sure you are locked in, and hang on tight, while we accelerate to our cruising speed. This will take about ten minutes."

Two minutes later, Byron was back on the speaker, "Hey, everyone, take a look out of the starboard side of the ship. Sorry, starboard means the right side of the ship. Our captain has modified the course so you can have a view of the NSEA campus.

In another minute or two the captain will roll the ship over so those of you on the port side – that's the left side - can have a look."

Several voices responded, "Ohhh , I think I'll be sick." "I'm not going to like this." "My eyes are closed."

Byron, on the speaker, "Now, everyone, calm down. This is all controlled flight, and you should not feel any different from riding in your car."

In another few minutes, Byron spoke again, "Ladies and gentlemen, we are now at our planned cruising speed, and we will be holding that speed for about an hour. That is, cruising at eight thousand miles an hour. The captain said you will be able to unlock your seat belts in a few minutes. Right now, he is maneuvering the ship to place Earth's gravity on the floor so you can get up and stretch your legs."

Shortly afterwards we heard the buzzer and then the Seat Belt sign went off. A few stood up, looked around and went for the head. It was like everyone came to life. Some are standing up, leaning over and talking with new acquaintances.

An hour later, we saw a new message, "Seat Belts locked in 10 minutes" Byron was again on the speaker and told us our ETA was 45 minutes. We felt several course changes and other unusual sensations as the shuttle was maneuvered to land at the designated position on Columbia.

Eventually we felt a thump-thump, similar to that when an airliner's wheels touch down on the runway. The difference is this is a vertical landing - like a helicopter makes. That was sort of a signal, and several passengers shouted out, "We have landed." A few minutes later, we heard workers on the outside of the ship securing the tie-downs. Then one of the shuttle's crew members arrived carrying some tools. He started working on the floor in the middle of the seating area. He then pushed a button and one edge of that section of the floor tilted up and we saw a set of stairs going down to Columbia.

Byron told us that the shuttle, Columbia, and SC12 all have oxygenated atmospheres, so we don't need to carry oxygen tanks. He also told us the gravity was somewhat less than we were used to. "Don't try seeing how high you can jump. You just might bounce off the ceiling and hurt yourself."

Still inside the shuttle, Byron told us to go claim our luggage, and then we followed him down the stairs and into Columbia. Several minutes later, Byron informed us, "You are standing in a small waiting room of Columbia. Escorts will take you through the underground to an elevator going up and inside SC12, leading to our new rooms, which we refer to as bunkers." Byron took another ten minutes to mention our PODs and our high-tech audio-visual-computer equipment that was already positioned in each of the bunkers. He informed us that the audio-visual-computer equipment is normally referred to as the AVC. Byron also reminded us the second group would be arriving late morning tomorrow.

With that, we were divided into four groups, and our four escorts, in turn, took us to the elevator that took us from inside Columbia to inside SC12. Our escorts walked us to each of our bunkers, and then would double-back a few minutes later to answer any quick questions we might have. Alone, in our bunkers, we had plenty to do. Among other things, we found a schedule for the first week. The instruction manuals for the PODs and the AVC appeared at first to be overwhelming, but after a few minutes we decided this was not so difficult and would be very interesting - as well as very necessary.

Settling In. Ray Hilt just said “thank you” to Byron Williams for taking him to his room. The door closed, and then he heard a slight buzz, followed by a voice saying, “Welcome aboard, Ray.”

Ray was confused. Where was the voice coming from? Then he heard, “This is your AVC speaking. Your audio-visual-computer apparatus in case you forgot. I have been programmed to help you become familiar with your bunker and pass on some tips that you should find helpful.”

When he stepped into the room, there was a brief message on the screen. He saw the screen, read the message, and then ignored it. Now, when he heard the voice, he saw the screen change, and he thought to himself, “Wow, talk about high technology.” This message was about flying missiles. “That’s a good point; I don’t want any flying missiles in my room – I mean bunker.”

The next AVC message was a lot broader, with instructions about other things to do on our first day. “There is a handbook in the drawer under the AVC that the AVC said we must spend a few minutes this evening. The first page repeated the welcome aboard message and the message about “no flying missiles.” The next page talked about our attire. The instructions were clear: “Remove all the clothing you presently have on and place that in the same drawer with the briefcases, purses, and overnight bags. In the next drawer below are several sets of the preferred attire. This is what you will be wearing for the remainder of your stay on SC12.”

There were several pages telling us how to place our orders for meals. In short, we pulled up a menu page on our AVC, selected our desires and hit the enter button. Delivery was in a small package sent by way of vacuum tubes. These systems became commonplace in larger stores and on Navy ships in the early-twentieth century.

NSEA

Message to Raymond Hilt

Urgent – Do this now.

Gather up anything that is loose and store it in the drawer next to the door. If the ship needs to make a sudden turn, you don't want any flying missiles in your bunker.

There was a page titled "SCHEDULE." This page shows our schedule for the first week. Day one was the day of lift-off. At the end of this page was a reference to the complete schedule in the AVC. If we wanted, we could print selected pages and insert them into our handbook.

The next morning, at 11:45 we were all peering into our AVCs to watch the approach and landing of the second group, on board the shuttle. It was an exciting event and while watching the graceful maneuvers of the space shuttle we became even more convinced the events we are about to face have been well thought out and expertly executed.

We were told not to bother those in the second flight until after four PM, so to give them time to become acquainted with the basics of what they need to do. Lift-off from Columbia would be the following morning, time not yet determined.

More Background. NSEA has been planning this mission for two years. Interplanetary travel is a lot different from circling the world for a few months in a satellite, and then returning to land. There are many new parameters that must be considered. The lift-off and initial acceleration phases take a lot of fuel. During their earliest flights, they used external tanks and engines to accelerate to perhaps one-half our basic cruising velocity, then jettisoned those and switched to their internal, main engines. For a one-stop mission – that is, out and back with no stops in route – these initial phases could require one half of the fuel for the total mission.

A visit to another planet, however, requires two stops – with two lift-offs - and more fuel. They eventually learned they could reduce the fuel requirements considerably for the first lift-off and acceleration phases by staging from Columbia, where gravity is one-half of the gravity on Planet Earth. But there is nothing they could do to reduce the requirements for lift-off and acceleration for the return flight. In fact, the problem is more complicated because they must carry all that fuel for the entire trip from here to there. Further, that planet may be larger than ours, and the gravitational pull would then be more than ours.

Current-day power plants are all nuclear, and most of those problems no longer exist. Nuclear energy has been our primary power source for the past forty years. It is a much more efficient fuel, and our engines are many times more powerful than in the twentieth century. In addition to making our ships stronger, lighter, and more heat resistant, we have learned to use nuclear energy more safely and efficiently.

A breakthrough occurred in 2055 with the discovery of an economical means to break through the so-called light-speed barrier. We had been experimenting with means of controlling the nuclear reaction. Moderated fully, the nuclear reaction is a highly controllable heat source, much like the mechanism used in our electrical power plants starting in the mid-twentieth century. By this means, the nuclear reaction is used to heat water to boiling, and the steam drives the electro-turbines that generate electricity. At the other extreme, if the nuclear reaction is not moderated, the result may be a nuclear explosion, like those first used in WWII, resulting in the loss of several thousand lives, but also expediting the end of that war and undoubtedly saving many times that number of lives.

Our scientists did a lot of work, both on the drawing board and in the laboratory. This was becoming a major project, and about to pass beyond those technologies our scientists were experienced in, or even fully understood. The space transportation organizations joined with the military science departments to establish an approved research and development (R&D) program.

New theories had to be proven, and that involved a lot of trial and error. Our goal was to augment the normal power plant with a series of very small, slightly modulated nuclear reactions. The initial reactions were too powerful, resulting in explosions that destroyed the power plant mock-ups and much of the surrounding equipment and apparatus for controlling the process and observing the results. After backing off to about one hundredth of that used in the initial attempts, they achieved a satisfactory event. With a few more trials, the process was refined to the point where they were ready to announce their success to the rest of the world. The available power for lift-off had then been increased by at least a thousand times!

But that was nothing compared to the next discovery. Using nuclear power for the lift-off had proved to be very beneficial. The scientists then focused on using nuclear energy for the acceleration phase. The normal Newtonian laws of physics were followed, and then Einstein's theories of relativity started to be evidenced as speeds then reached above 90 percent of the speed of light, abbreviated L0.9. Modest increases in power seemed to have little or no effect on speed, so eventually someone decided to double the power. The results were positive, with maximum speed then up to L0.95. The scientists were delighted. In the next experiment, power was again doubled, but the resulting increase was much smaller, reaching a speed of about L0.96. Several more procedures called for doubling the power, and the observed results were hardly discernable.

Eventually, the trial power was up to fifty times the amount that resulted in L0.99, and history was made. They had broken through the light barrier, reaching L1.05. Was Einstein wrong? The scientists finally concluded that there must be three sets of laws pertaining to motion physics. Newton's laws apply to the "normal" range of speeds, up to about L0.8. Einstein's laws apply to the "transitional" range, about L0.8 to L1.0. Some other laws, then, must apply to the higher speeds. The scientists then established a high priority to investigate this further, but our space business had been keeping them so busy, they have not made any significant progress with that research.

After we have reached outer space, we are in a total vacuum, and we have no gravity. These are wonderful conditions for space flight; we then shut down the engines and the spaceship continues at a steady speed. At these speeds it takes as long to decelerate as it took to speed up in the first place. When we want to slow down, we can't just fire the engines in reverse – we don't have a reverse; so, we have to "flip" the cruiser, nose for tail. As it turns out, flipping positions puts our artificial gravity where the floor is, and gives the passengers a more normal feel.

The normal laws of physics govern these relationships. In these conditions, we can reach speeds that are equal to whatever is leaving the thrust pipes of our power plant. Theoretically, after we are in the range where Einstein's laws apply, we could expect it possible to accelerate to the speed of light by shining a flashlight out the back of the spaceship. But normal light has no mass, and both Newton's laws and Einstein's laws say we need to exert a force out the back for the spaceship to accelerate forward with the "equal and opposite reaction." Photons, however, do have a mass – or the equivalent of a mass. Granted, these are very small, but this may help to accomplish another major step for space travels.

Then the scientists and engineers had an idea for an engine that was more efficient than using small nuclear explosions. This involved studying up on some of the more advanced fields, such as quantum physics, electromagnetic radiation, particle physics, and several others. After an intense engineering effort that lasted several years, a usable method was refined, and it became a very successful propulsion mechanism for the next ten years.

During that same period a group of engineers and scientists continued looking for other ways to break through the light barrier. A breakthrough occurred when we discovered that photons can be accelerated several-fold by using lenses to condense or compress the photons. This is somewhat analogous to the Bernoulli effect, observed when the speed of water going through a pipe will increase when it reaches a section of pipe with a smaller diameter.

One facet of Einstein's laws proved to be very useful. As the photons are accelerated, the mass increases. This, then, was the basis for our next-generation propulsion engines. But the increased mass of the photon is still very small, and therefore our acceleration is quite small, so it takes a long time to reach several multiples of light speed. We have used spaceships for two years now that have reached L2 for recent explorations, and we may be using a cruise speed of L8 on this mission.

Lift-Off. Launching from a space station is different from an Earth launch. SC12 is currently landed on and secured to Columbia, which is sometimes referred to as the "Iron City in the Sky." As the scheduled time approaches for lift-off, Captain Nick Logan gives the signal to his engineering crew to start warming up the engines, and then orders more and more of the tie-downs released. The crew members confirmed their readiness and when the captain discerned all appeared to be right, he ordered the engines to Lift-Off power.

"Three…two…full power…NOW!" This is not the blast-off type that was the norm in the early part of the 21st century, although it is felt throughout the ship, and there is no doubt in anyone's mind about exactly what is happening. All eyes are glued to their AVC screens that are displaying live video from the external cameras. The sensation of Lift-Off is felt as a slight jerk, like a fast start in your car when the light changes from red to green. The big difference is that the acceleration does not stop. The feeling of acceleration is not great, slightly more than One G, but it is steady, and we can definitely feel the sensation of increasing speed. The picture on the computer screen is the same live video that is going out to the general public via the TV news casts. One big difference is that we're inside this beautiful and magnificent machine, and we can feel and hear everything that is happening. This is not a movie. This is not a drill. This is for real.

The launch was successful, and all went as planned. The Captain held the acceleration to 1.2G for a few minutes. A "G" is the equivalent of the same force of gravity we are familiar with on Planet Earth. As planned, after two minutes at 1.2G, the Operations Officer keyed the mike button and said, "Hello everyone, let there be no doubt. We are underway. When you hear me say NOW in about a minute, everyone should hang on to something while we do our acceleration test. We will be going up to three and a half Gs and remain at that level for about a minute. This is an operational "test" to confirm that the ship is capable of performing acceleration changes, which will be necessary from time to time during our mission. This will also give the new passengers a touch of what they will be experiencing again in another few days, but for longer periods of time."

We quickly learned that 3.5G is a lot different from 1.2G. After the test, the Captain ordered a fixed speed of 5,000 MPH for several hours. While cruising at a fixed speed, we varied between zero and one G. At one G of acceleration, we are oriented upright with our feet on the floors. At zero G we are floating.

Several hours later, as we were becoming familiar with cruising at a high speed, we started wondering when the captain would start the acceleration phase. Soon after that, we all heard a brief "beep" on the pri-comm, and we turned to our AVCs to see and hear what was going on.

"Hello everyone. I am Commander Castner, Operations Officer of SC12. I would like to have your attention for a few minutes. By the way, we are all on a first name basis on this ship. My first name is Bert and Captain Logan's first name is Captain, or Skipper. I'm sure you will get to know all of us very well by the time we arrive back home."

"And now, some ground rules. Our overall mission consists of three stages. Stage One takes us from home to our destination. Stage Two is our exploration and discovery mission on planet Alpha-1. And Stage Three will take us back home. Stage One consists of five phases: the launch phase, then acceleration, cruise, deceleration, and landing. Five phases. During our cruise, we will spend a good amount of time preparing for Stage Two, so we don't have to go into that right now. And then Stage Three is our return trip. This is a repeat of the same five phases as in Stage I, just going the other direction. Now, back to Stage One."

"We have just completed phase one, Lift-Off, and now we are preparing for the first part of phase two, acceleration. The early part is the easy part. At one G, you should feel quite comfortable. This is a good time to try out the different options on your POD. You need to find the position that is most comfortable for you, and then hit the memory button on your POD keypad. This way you won't have to spend time searching for it when you might need it in a hurry. Now, let's talk about some rules. During this one-G part, we require two straps while you are in your POD. One strap goes over your thighs and the other goes around the lower part of your chest. The purpose of the two straps is just to keep you where you want to be in the event of turbulence, or a rapid maneuver."

If you need to move around inside your bunker, you will always need at least a one-point connection. There is a travel lanyard attached to a loop on your POD. Leave one end fastened as it is and hook the other end to the loop at your belt. When you leave the POD, the lanyard should be long enough for you to reach anything in your bunker. When you want to leave your bunker, go to your door, string the lanyard through your belt loop and connect to the loop on the wall by the door. Then disconnect from the loop at the POD, step outside the door and connect that end to the loop next to the door. Then disconnect from the inside loop, step back outside and look for the next loop in the direction you want to go."

"Everyone needs to remain cautious about gravitational pull. If you step out of your bunker with nothing attached, you could fall and hurt yourself. If you were in a hotel, on the third floor, and stepped out of an open window, you would probably be killed. When SC12 is flying at one G, the "pull" is the same as on Earth. If we are at three Gs, the pull is three times stronger. We can't emphasize this enough. You want to be attached to something at all times."

We could see others trying to mimic the motions they would soon become familiar with. That was a good sign.

Cdr. Castner continues, "When you leave your bunker, you must be positively connected to something at all times. And you must key in your destination in your AVC before you leave. For example, when you want to go the head, the first step is to select the "head" button on your console. This tells the watch officer you are going to the head. In case the head is not available at this time – it might be in use, or it might be inoperative for some reason – the watch officer will notify you, and the two of you can discuss alternatives in case you have an urgent need. If you are headed for another bunker, a meeting room, or a workshop, go to the "TRAVELING" page and key in your destination. If you are staying on the same floor, the AVC will most likely answer, "OK." Trips to a different floor will require a ride on the elevator. If there is no problem, the AVC will answer, "ELEVATOR," and your trip is approved."

Cdr. Castner went on to talk about the normal daily routine and did a good job of covering many of the unexciting things, like keeping track of personal items and storing everything in the drawers so they can't become flying missiles. Baths and showers don't exist on space cruisers, and personal hygiene is taken care of by occasional visits to the ultrasonic closet, where a thirty-second treatment does an excellent job of removing anything that is not firmly attached and provides a gentle body massage at the same time."

He ended his talk at that point and informed the listeners he would plan to return for another fifteen-minute session tomorrow morning. The next morning Cdr. Castner continued, "Good morning, this is Bert again. I have another page of notes that I need to finish, and then I'll turn the microphone over to the Captain."

"While we are accelerating at a one-G level, things will appear to be very normal. On Earth, we were always in a one-G situation, caused by a constant force of gravity. We just naturally "stand" with our feet toward the center of the earth. We have spent our lifetimes getting used to that."

"We will have periods that may feel similar to that on Earth. But a one-G situation in a space cruiser can be an entirely different matter. We are distant enough from planets and stars that we do not have a gravitational "pull" from them. The one-G situation is created by adjusting our throttles to achieve one G of acceleration. When we reach our desired speed, we can shut down the engines and continue to speed along at the same speed. There is no friction in outer space, and we expect to be far enough from other planets and stars that our ship will continue in the same direction and at the same speed. That is one of Isaac Newton's laws. Our feet just naturally stand on the floor, and our heads are toward the pointed end of the ship. Eventually we will arrive at our cruising speed. We are as fast as we want to go, or need to go, and the Captain can shut down the engines and we will continue to "cruise" at that same speed. We call this the cruise phase. There is no feeling of acceleration, and we can "float" from place to place."

At this point, Cdr. Castner shifted to the subject of bunkers. "Just so you know, we have two hundred and forty bunkers on SC12. Two hundred of these are designed for one passenger each, and the other forty are for two each. One of our bunkers is currently assigned to a married couple, two are for siblings, and several more are for good friends. NSEA officers, ranked lieutenant and above, are in single bunkers. Lower ranked NSEA personnel all have roommates. Each bunker is equipped with one or two "PODs." A POD serves as the chair where a person sits during the awake periods and also lays back in a horizontal or reclining position at other times. PODs are manufactured in several sizes and are also adjustable to fit whatever comfort requirement people may desire,"

The next section of Cdr. Castner's talk is about the AVCs, a benefit that the academic types found very appealing. "In front of each POD is a very high-technology audio-visual-computer setup; reachable within the system is an electronic library of just about every book published, including novels, textbooks, and much more. One can register for academic courses that can be taken, and then earn certificates that can be presented for credits to the college of their choice at a later date. If one needs a break from educational aspirations, there is a fine selection of movies. Also available are many computer games."

Other attributes were discussed. "There are several medium size rooms for group gatherings, or meetings, and two exercise rooms. We have a rather large garden for growing vegetables and herbs. There is a kitchen, of sorts, although some have described this as more of a chemistry lab. Passengers can order their meals from their PODs. The kitchen prepares the meals which are then delivered the following day.

Meals are artificial of course, but are full of vitamins and nutrients, and include a few popular flavors of such things as grilled hamburger, broiled chicken, bacon, fish, and so on. Meals are available in either liquid form or chewable solids. There is a recycle center that recycles everything. Yes, that includes everything. Scientists and engineers have reviewed the overall setup and assured us the ship's recycle system can support a complete load of crew and passengers for more than a thousand years."

Commander Castner then said, "Now it is time for me to introduce your Captain. Captain Logan is undoubtedly the most experienced Cruiser Captain in the Space Fleet today, having completed seventeen space missions, including three missions as Captain, and each with more than one month in space. His most significant achievement was the famous mission to Pluto, which earned achievement medals for the entire crew, and as well for many of the scientists for the new geological discoveries from that mission. He also distinguished himself as a Test Pilot for several experimental ships and was the Senior Test Pilot for this ship."

At this point, the Captain took the microphone and started his talk, "Thank you, Bert, and Good morning, everyone. I only need your attention for a few minutes. First, I am delighted to inform you that we have successfully completed the first phase of our mission. The break-away and lift-off went very well.

All of our flight instruments look good, and all systems are ‘go’ for continuing with our mission. Just in case you may have forgotten, our estimated time to return is about two years, plus or minus a month or two. I know you have all been briefed on our schedule, but I want to review several points with you, and make sure everyone understands the ground rules.”

“We are presently cruising at 50,000 miles per hour. We plan to start phase 2 this afternoon, which will involve accelerating at one G. This is a very comfortable level, and we plan to continue at one G for two more days. By the end of those two days, our speed should be about three million miles per hour. Then we will start the second part of the acceleration phase and things will get a little more complicated. We will be slowly increasing our acceleration rate, going up to two Gs for two days, and then four Gs, and so on until we reach nine-tenths of light speed (L0.9).”

“Your bunkers are outfitted with special G-suits that provide support to our sensitive body parts and help us keep from passing out when we are in higher G situations. Note that all of us will be required to have our G-suits on before we go higher than two Gs. We plan to continue at four Gs for two days. Then we will give you plenty of notice to select a higher setting on your G-suits. Then we will all need to hang on tight for the next part when we gradually increase to eight Gs. Your PODs are built to provide you with additional body support for this part. Also, you will find that our bodies adjust to these things quickly, and the acceleration forces become more tolerable after a short period. We will continue stepping up the acceleration, and, if all are willing and capable, we may eventually go up to ten Gs. Our scientists have assured me that our equipment will safely provide the protection we will need for ten Gs. This will take us one to two months, depending on how much punishment we can tolerate."

"We are aiming for a cruising speed of at least L8. Much less than that and we might be old people by the time we get back home. Actually, there is a pretty good chance that we can go even faster, but I'm not interested in setting records. My job is to complete this mission as planned, and to return home with everyone we started out with. I have to go now. We will talk more in the next few days."

Day 5. Acceleration. Two days later Captain Logan was back on the microphone. In his nice, calm tone, he sounds like a football coach helping his players understand how to work a complicated play, "Now remember, we gave you a taste of acceleration the first day out. Right after lift-off we adjusted the throttle and continued at one G for several minutes. Then we increased our thrust and went to three Gs for a brief period. That step was necessary by our operations manual and confirms the engines are capable of higher power maneuvering when we need it. That's about it for now. Bert will be talking with you shortly, and then I'll be back on the microphone sometime tomorrow."

An hour later, the buzzer sounded, and then we heard, "Good afternoon, this is Bert speaking. We are currently at zero G, cruising along at a steady speed of 50,000 miles per hour. We have been doing this speed for six hours now, and the captain is preparing to shift the throttles to a moderate acceleration of one G. Remember, one G is the same as what we used on Planet Earth. One G enables us to walk around on the normal surface. Think about a multi-level building. We are on the first level, and someone calls us to a window. That 'someone' wants to chat with us on the ground level, and we jump out the window. This is only a few feet to the ground, and we can jump out the window without injury.

"Now think about this. If we jump from the tenth floor, we will start at the same speed, but then our speed would increase significantly and we would most likely die from the fall. Why does the fall end up so fast? The reason is the gravity on our body causes an acceleration and as long as we are "falling," the speed increases.

"You were only in the air a few seconds – maybe half a minute - , and you accelerated to a hundred miles per hour. Well, our ship is going to keep accelerating. Remember what the Captain said a while ago, two days from now we'll be up to three million miles per hour. Three days later, that should be in the billions."

A few minutes later, Bert was preparing to increase our speed. "My AVC shows that all of you have your safety belts fastened, here goes the three-second warning to go to one G. Three…Two… One…Execute!"

"Now, let's accelerate. The ship is positioned with the pointed end up forward, and our feet will be walking on the floors, so we should be quite comfortable,

We have adjusted the engines to maintain our acceleration at one G. Our current speed is about five thousand MPH, and tomorrow morning we should be at six or seven-hundred thousand.

If we were in a hurry, we could have selected two Gs and we would only need a half day to reach twenty thousand. Well, how about ten Gs? Or twenty Gs? We would find it very difficult to move around in our bunkers at four or five Gs. Most of us would pass out at the five or six G level. At only a few more Gs, it would be impossible to breath, and we would soon die from asphyxiation. These are things that we must be very careful about. And that's all I have for now. More later."

Day 6. 2.5 Gs. At 9:00 the morning of day six, Bert was back on the pri-comm, "Good morning, everyone. As we discussed yesterday, we will be increasing to two and a half Gs. This will be at slow intervals, so you can become used to the changes as we proceed. We will have our first increase in about five minutes, and you must press your "ready" button on your AVCs. Just a few seconds before the start time, we heard over the pri-comm, "This is Commander Castner. We have three people who have not responded on the READY button. You are holding up an important event."

One of our team started donning his G-suit a few minutes after 9:00. He then pulled up a book on his AVC and commenced reading. Apparently the first chapter was not very exciting, because he fell asleep. At a few minutes before 9:00. there was a buzz on his AVC. His sleep was deep enough he did not hear Bert's request. At 9:05 he was finally awakened to a loud horn from his AVC, and the screen was flashing, "Put on your G-suit, strap in on your POD and press "READY." He finally did as told, and the horn was quiet.

Our G-suits are advanced versions of the G-suits worn by military tactical aviators in the mid-twentieth century. Those G-suits included a rubber bladder in front of the abdominal area, and the large muscles in the upper and lower legs. When the pilot was maneuvering his aircraft into a steep dive for a bombing run - or struggling to be the winner in a one-on-one shoot-out with an enemy fighter - he may need to pull five or six Gs for a short period to make a tight turn. Without a G-suit, a pilot could black out or faint if the "pull" goes longer than his tolerance level. Captain Logan is being very careful to help us advance our tolerance levels. The G-suit will increase our tolerance by at least several Gs. But spending several hours at a particular G-level will eventually increase our tolerance even more than the G-suit.

Our new G-suits cover the whole body, with rubber bladders in the appropriate places, and a polycarbonate or acrylic helmet. The suit plugs into the POD, which provides compressed air according to the need.

Time to Relax. Friendships are being formed. Mary Smith found need of a sick bay. The hard way. She twisted an ankle while returning to her bunker. She was limping, favoring the injured leg, and looked up "sick bay" in the handbook. The ship had a very small medical staff. One MD and one Medic (similar to a PA). She keyed the number for sick bay. Two rings, and then a voice, "Leave a message and I'll call back." She left a message, tried to make it sound like she was writhing in pain, otherwise she might never have a reply. Two minutes later she heard a knock on her door, and a voice, "This is Charlie, the Medic, answering your call.

She answered quickly, "Come in."

He entered and said, “I was on the phone when you called. Sick Bay just happens to be on the same floor as your bunker, so this should save time for both of us.” Charlie was polite and very professional. He provided a Tylenol for pain, rubbed the sore area with a salve, and put on an elastic wrap. He smiled and said, “You should stay off that leg today. I’ll stop by in the morning for a follow-up. Gotta go. See you later.” A friendship was born.

Larry Bailey spent some time browsing through some courses that were available. Larry had a BS in electrical engineering, and an MS in physics, with a space traveling specialty. He had purchased several telescopes and had thought about building one of his own design. It was interesting that when he started reviewing courses available, names of personnel on the ship with similar education and interests were identified. Larry saw the name, Dick Alderman, with a BS in math, and an MS in mechanical engineering with a specialty in optics. Larry keyed Dick’s number, and they ended up talking for an hour. They sounded like two lost brothers and decided to get together the following day. Another friendship.

Elite guests. Hank and Betty were the only husband-wife team on the ship. They met in college when they were both physics majors at the University of Maryland. Hank was a senior for his BS degree and Betty was in the doctoral program. Hank had earlier left college after his second year to enlist in the Army and returned four years later to continue his education. He then met Betty while she was working part-time as a lab assistant and helping the undergrads with some of the harder parts of the program. A romance developed, and they were married right after Hank earned his BS degree, and Betty received her PhD at the same time. At the end of their service, the minister said "…and now I would like to introduce Mister and Doctor Hodge." Most of the wedding guests were college contemporaries and faculty, and they really got a kick out of that. Hank finished his MS degree two years later and decided that was enough education for him.

Both Hank and Betty ended up working in the space exploration program at the JHU Applied Physics Laboratory near Laurel, Maryland. They both proved to be highly skilled research scientists, although they had different specialties and different styles. Betty chose the theoretical side of physics, and Hank stayed on the more comfortable side, focusing on practical applications. Some of their joint presentations in the academic circles were legendary, especially the one where Betty presented her theory that the light speed barrier could be broken by using a series of photon pulses with increasingly large amounts of thrust. Hank followed her part with his descriptions of exactly how this could be achieved with carefully controlled applications of quantum physics and nuclear power. It was this research project that sparked the space-travel community to concentrate on achieving what is now called Light-plus, or L+, speed.

This is the first major excursion outside of our solar system, and Hank and Betty were invited to join as special consultants. This was easy to do since they had already become personal friends with Captain Logan. Although their research had marked the real beginning of the L+ development and the creation of the Propulsion Division of NSEA, their professional interests eventually moved on to other areas. But they continued to monitor the L+ program, just as one would proudly follow the achievements of a son or daughter as they mature and become successful in their professional fields. Now in their middle thirties, Hank and Betty knew that their invitation was really honorary. They had no assigned responsibilities, other than to attend "as they may desire" any of the technical or operational meetings.

Little did they know – how could anyone know

that their talents would again become so valuable.

Hank and Betty were not members of the flight crew, and had been assigned a workstation bunker, just like most of the others on the ship. Designed for two people, their utilitarian bunker was a combination of living and working space and contained two AVCs with active links to the main network. Hank and Betty were provided full access, albeit read-only, to flight control and mission information. This allowed them to monitor just about anything they chose without the ability to actually change anything. They had unlimited use of the vast digital libraries which contained impressive amounts of research material and technical data that they expected to make good use of during the two-year trip. They also had access to a wide variety of virtual entertainment, including books, movies, music, and computer games. After all, they would have a lot of time on their hands.

Error in the Instructions. Around the middle of the second week, we were at a speed of L0.9, almost light speed, and all had gone well. The procedure then called for a big jump in power, in order to break through the light-speed barrier. Captain Logan had pulled up the detailed directions to re-read the important part: "Schedule power for ten percent increase; double check all settings; press EXECUTE. You will feel a rapid acceleration. After five seconds, press the same button for OFF." It seemed so easy. But nothing happened. What went wrong? He brought in Bert to assist with the procedure. About midnight, they found the latest draft version of the procedures, and decided to compare that to the approved version.

Bert became temporarily engrossed in the draft, then: "Ah-ha! I found something." He set the book down in front of Nick and continued, "Apparently, the authors were not in complete agreement." We could see several cross-outs. It was impossible to tell if the correct number was ten percent or twenty percent.

Nick and Bert were excited, and they wanted to try again; now, at 2:00 in the morning, Nick hit the pri-comm button and said, "Ladies and gentlemen, I am sorry to interrupt your sleep. We have been struggling to push SC12 through the light-speed barrier, and I think we have solved the problem. You will feel a jerk in a few minutes, and that should be it. But if it doesn't work on the first try, we might try another time or two. That is all. And good night."

Nick set the increase to twenty percent and Bert hit the execute button. Nothing. Nick turned it up to thirty percent. Bert looked at Nick. Nick nodded his head. Bert pushed the button. Nothing. They tried it one more time, at forty percent. Nothing. They gave up and decided to try again in the morning. Nothing. What can we do?

Hank and Betty had been involved in setting up some of the operational procedures for going through the several complicated speed maneuvers.

Hank dialed the Captain's private number, "Nick, Betty and I might be able to help. Can we come up?"

Captain Logan replied, "Yes. I had forgotten that the two of you were involved in developing these procedures. Come on up. The door is unlocked. And thank you for calling."

Hank and Betty took the elevator to the "top." Hank tapped on the door and announced, "It's Hank and Betty," and Bert opened the door.

Nick and Bert had spread out the applicable documents on a table, and Hank was reviewing them. He put his finger on the page with the scribbling on it, and said, "Betty, Look here. You should recognize this."

Betty needed only a few seconds, and she said, "There was a big argument about semantics. We agreed the correct step was to double the thrust. I wanted the manual to say it that way. "Double the thrust." That leaves no doubt about what to do. But, no, they wanted this to be "more precise." One person said to increase to two hundred percent. The other said, no, increase it by one hundred percent. I told them that those were confusing terms because nothing preceding this was specified in percentages."

She pointed to the pencil marks and said, "See where it says double, Well, that is my writing. You can barely see where it says 100%, and 200%, but the technical typist erroneously typed it as 10%. I heard of this a few days later and reported the error. Obviously, it was never corrected."

Nick added this, "You know, I had started increasing the percentages from ten percent to twenty, then thirty. But I never would have gone so far as to double the thrust."

Betty replied, "Yes that is a big jump, but that is correct, and it is only for five seconds."

Nick thanked Hank and Betty for their help on this, and said, "Why don't you stay with us for a few minutes, and we will try again.

Bert turned on the pri-comm. Nick grinned and said into the microphone, "Ladies and Gentlemen, this is your Captain speaking. We just had some very professional help on our situation, and we are about to try again. I would like for everyone to be strapped in as we do this. In five minutes."

Five minutes later, Bert was on the pri-comm, "Attention everyone. Stand by. Three … Two … One … Now." Nick had made the setting for one hundred percent increase, and he nodded his head to Bert. Bert punched the Execute button. Everyone in the ship felt the jolt as the engines doubled the thrust. Five seconds later, the speed indicator jumped from L0.9 to L1.2. Bert yelled, "Yahoo," and punched the button again for OFF. Nick, Bert, Hank, and Betty hugged and shook hands to congratulate each other. Nick keyed the pri-comm Button and said to everyone, "Well, we made it through the light-speed barrier. The controls are now set for one G acceleration. Bert and I will work out a new schedule and we will bring you up to date in the morning. Good night again."

Nick looked exhausted, and he said to Bert, "I would guess this set our schedule back several days. Tomorrow morning you can help me figure out a new schedule to bring our acceleration back to normal. Then we can bring everyone else up to date and give them an alert for G-suits. You get some rest, and I'll see you in the morning."

Day 10 is a busy day. Captain Logan talked to us at 10:00 about today's schedule. It was important that we gain more experience at the higher G level, so he started leading us through the range of Gs for today. We moved up to 1.5 Gs at 10:00, and 2.0 Gs at 11:00. We were prompted to don our G-suits a few minutes before 11:30, and then at 11:30 sharp he increased our acceleration to 2.5 Gs. We remained at 2.5 Gs for a full hour and then he increased to 3.0 Gs at 12:30 and 3.5 at 1:00. We remained at 3.5 Gs until 4:00.

Later in the evening, Bert talked to us on the pri-comm, "Captain Logan told me he is backing off to 2 Gs for the rest of the day, and until tomorrow morning at 9:00. The new acceleration schedule will put us at 3.5 Gs at 9:30. Then we will level off at 4 Gs about 1:00 and we will continue at that level until 4:00. Continuing this schedule, we should be up to 8 Gs in another week. By that time our speed should be up to L5 or L6, and then the Captain said we would coast for a few days. After our rest period, we will have a new schedule to go up to L8. That's all for now. Either Captain Logan or I will be talking to you each morning."

Sunday Church Service. This subject appeared on the schedule in our third week: "Those interested are invited to a meeting next Sunday. The purpose of the meeting is to determine if we should have Sunday services on a regular basis. Details will be worked out at the meeting."

There were thirty people at the meeting, Bert opened the meeting, "Today's meeting is not a church service. The purpose of this meeting is to decide if the number of interested people is sufficient to warrant reserving the room on a regular schedule, and perhaps supporting any specific needs during our trip."

We had a very good interest for two services. The leaders agreed to services every Sunday, with Catholic services at 9:00 and Protestant services at 10:30. At one point in the meeting, Bert asked for a show of hands for those who wanted Catholic services, and a few minutes later for Protestant. After the count was taken one man stood up and said, "I am neither of those. I am Jewish and I was hoping there would be more of my faith, but that didn't happen. I would like to sit in for the Protestant services - if there are no objections. Will that be OK?

The nominal Protestant leader grinned and said, "No objection. We'll take anyone, just so we can keep up with the Catholics."

The next Sunday saw about forty people for each of the services. We didn't have ordained ministers, of course, but we managed to sign up some good volunteers. Thirty-minute sermons were outlawed. Instead, we would pick two or three volunteers to give their "church stories" for the following Sundays. Each congregation elected their own church leaders, who were expected to fill the position for several months, unless requested to resign earlier.

Zero G. Captain Logan was a good leader. He worked his staff members quite hard and enforced the ship's rules on the invited guests. They all responded in a good fashion. Then it became time for two days of zero G. All of us had experienced zero G several times, and it was fun to float through the air.

Free time in the bunkers was more enjoyable. Straps or lanyards were not required, although they were still recommended. Two straps were still required when we turned in for the night.

There was a small basketball court (one basket) next to the physical fitness area, and several groups had been forming basketball teams and challenging others to a game or two. Those events had been during one-G periods, and now they looked forward to zero G so they could see what that was like. Their favorite tactic became dribbling to the far end, turning around and then, with basketball in hand, leap for the basket and try to score. The free throws were particularly interesting. Instead of throwing the ball, they would leap toward the basket and then sink the ball when they were close to the basket. They sort of invented a new game. We called it 3D Basketball.

Day 24. More Problems. Our period of fun and games was over. Zero G was replaced with 2 Gs, and our acceleration schedule moved us quickly to 4 Gs. A new schedule was implemented the following day that would take us to 8 Gs on Day 29. Then we would be rewarded with another two days at zero Gs.

Day 25, when we reached 5.2 Gs, we all heard a very loud noise. **Heavy vibrations** were felt throughout the ship. The Captain immediately backed off the power to bring us down to 5 Gs, and the buffeting stopped. He turned to Bert, hoping for an idea to appear. Nothing. He waited for an hour, collecting his thoughts and trying to figure out what happened. Nothing came to mind, so he started increasing thrust to take us back up to 6.5 Gs. This time he was very cautious. They reached 5.1 Gs in an hour, and 5.2 in another. Just a minute after that, the noise and vibrations started again. The ship remained at 5.2 Gs for two or three minutes, then Nick slowly backed off some of the thrust. Thinking to himself, but muttering out loud, "Maybe we can still complete the mission. Accelerating at a lower G level means it would take longer to reach our planned cruising speed of L8." Another minute later: "There, the buffcting stopped at 4.8 Gs. Let's stay at 4.8 and see how long it takes us to get the speed back to L3.7."

Nick decided it was time to share the problem with headquarters. Communications with NSEA headquarters was no longer a simple matter. Muttering to himself again; we are about one-fifth of a light year away from home. If our message had to travel at the speed of light, it would take one fifth of a year for a message to reach NSEA. Two and a half months. Fortunately, the communications technicians were able to install some fantastic updates and we are now able to send and receive messages at twenty times the speed of light. L20. Wow, maybe one of these days our space cruisers can match that. At L20 a message to NSEA had a travel time of just under four days. Each way.

Nick had been keeping Bert up to date on our problem, and he asked Bert to read the message he was preparing. Bert took a minute to read, and another minute to digest everything. Bert had been there with Nick when the problem first happened. Now, he smiled and said to Nick, "That looks good, boss. I can't think of anything to add." They worked well together, but this was apparently not a productive time for ideas. Bert looked toward Nick and said, "Is it time for us to call Hank and Betty?

Nick answered, "Yes that's a great idea, but I have to discuss with HQ before I can bring in Hank and Betty. They slapped my hands with the other problem. They don't want to read about a problem in a weekly report. They want to know right away. You can call Hank and Betty right after we send our message to NSEA."

Bert grinned, and said in a low voice, "Got it."

Nick pressed the send button. His message to NSEA was not lengthy, but with enough detail so the engineers could understand the problem, and then hopefully come up with a method for correcting the problem - or avoiding the problem. It would take seven or eight days for the messages to go back and forth. Add to that a day or two for the engineers and scientists to put together their recommendations for Nick and Bert to go through. Nick made a mental note to start looking for a reply eight or nine days from now.

Nick and Bert continued to "experiment" with their procedures, hoping to find a way around the problem. This resulted in one or two periods of buffeting, just about every day.

Two days later, several floors below the cockpit, "Hank, I've had it! If this shaking doesn't stop, I'm going to scream!" Frustration and agony were clearly visible in her face.

They were both strapped down tight, but Hank could see that she really hated this. "Hang in there, Betty. They'll figure it out soon."

"Yeah, I hope so. But we've been going through the same routine just about every day. She took a deep breath, then let the air out slowly, trying to calm herself. "I wish I could drop out of this mission. I give up. Get me out of here!"

"Oh, come on. You're tough enough to handle this." He knew from past experience that she could take it. Like the time when her professor told her to try a different approach on her thesis. She had to trash a year's worth of research and start over. A lesser person would have given up. But not her, she toughed it out and handled it just fine. He knew she could handle this too.

The next day, Nick called Hank and Betty's number, "Do you suppose the two of you could come upstairs to give us a hand?"

Betty replied, "Of course we can. We'll be right up."

Bert raised an eyebrow, "Didn't you say you had to talk to HQ first?"

Nick smiled and replied, "If they come up with a solution, I'll tell them you already thought of that."

Bert said, "Thanks, maybe I'll even get a promotion."

Nick and Bert spent several hours describing the problem, what made the buffeting start, and what they did to make it stop. The four of them spent the next several days reviewing the technical manuals, hoping to find a solution.

Day 27. Zero G. The next morning, on pri-comm, "Good morning, this is your Captain speaking. Yes, we have a problem in the upper G range of acceleration. We sent a report to the engineers at NSEA, and I hope to receive some suggestions to take care of the problem. We will push the problem aside for a few days, and everyone can enjoy a Zero G vacation. This is Day 27, and we will plan to return to a new acceleration schedule the morning of Day 30. I think we can start at 2 Gs, with a short warm-up so we can quickly move up to 4.5 Gs."

"By the way, Bert has been telling me about this new sport you invented, 3D Basketball. I asked the electricians to install several cameras there, so the 3D players will now have a larger audience. Bert will be touching base with you to learn about your next competition schedule. Then we can pipe the schedule into the AVC. That's all I have for now. Bert and I will keep you up to date on everything for the Zero G period and the status of our problem after we receive some recommendations from headquarters."

3D Basketball. Basketball was suddenly becoming the most important thing in the world – no, in the universe! When there was a game in progress, spectators filled up all the remaining space. And then when thc game was approaching the end, players for the next teams were standing in the halls, ready to squeeze in when the other players were trying to squeeze out. Spectators developed a strong loyalty to their favorites. Players and spectators alike were begging Bert for another basketball court.

Bert wondered if one of the other meeting rooms could be modified to be a second basketball court. He discussed the matter with Captain Logan. They started checking the backgrounds of their engineers and scientists, hoping to find someone with experience that might help. They hit pay-dirt right away, shaking and buffeting.

Jerry Rumble has an MS in nuclear power engineering, but his hobby is renovating older homes. Paul Webber is a senior electrical and electronics engineer, with plenty of previous experience building and wiring new homes.

Bert took the two engineers to the meeting room. Then he took them to the tool room, where they would find everything they might need. And more, “This is the spare parts room. Over there is a stack of spare walls. You see, most of the interior ‘walls’ in SC12 are movable, or you could say, relocatable. We could use these to convert the extra meeting rooms into smaller rooms, or we could remove some walls to convert several smaller rooms into another,”

Pause, and then “Another what?”

Bert looked at the two engineers, waiting. Finally, they caught on, and both said, in unison, “Another basketball court!”

“Right,”

They took the elevator back to the top to visit with Captain Logan. He started the conversation, “Well guys, can you help us with this?” They both nodded their heads.

Jerry was the first to answer, “Yes, I can help. Truth of the matter is, I would rather work on renovating anyway.”

Paul’s head was nodding again, “This sounds like fun. Yes, I want to help. When can we start?”

Captain Logan answered, “Ok, the project is approved. Bert is in charge. The two of you keep Bert informed and Bert can give me updates. I thank you both for helping with this. I’m sure the increased morale will be worth your efforts.”

An hour later, Bert returned to give his first update to Nick, “You may find this difficult to believe. They will be finished tomorrow afternoon. I think the whole ship is like a Broadway stage, with everything removable, replaceable, and relocatable.”

The next morning Bert made a detour to check on Jerry and Paul. Bert saw there were two new faces. Paul said, “Bert, this is Doctor Tim Wilson and that is

Doctor Lee Stieber. They are retired professors from Johns Hopkins University. They taught me the fine arts of electricity. Now I'm teaching them how to move walls."

Everyone shook hands. Jerry stepped over and said, "We should be done in about two hours." Bert was anxious to give his new update to Captain Logan.

Days 25 to 35. Looking for a Solution. Nick, Bert, Hank, and Betty reconvened just about every day. All four of them were nearly experts on the subject, and they were confident they would find a solution before they received the reply from NSEA.

Day 28. After two days they figured it was time for a different approach. Betty came up with an idea, "Let's make a list of things that could possibly cause the buffeting. Then we can go down the list, carefully, and discuss the possibility of a particular failure causing this particular problem. We are looking for cause-and-effect relationships."

Making the list took a little more than an hour. Discussing the first item took twice that. The list included forty items.

The morning of Day 31, they gathered in the cockpit. They had long faces because they had drawn a blank on this. Betty opened, " How about another approach. Let's pretend we are designing a new ship, or at least a new system that might be involved here. Then we can discuss our design and look for any weak spots, or any suspicious parts that could be causing our problem."

Nick commented, "That sounds like a good idea. Betty, you are good with theories, so why don't you be the designer. The rest of us can take the practical approach while we're looking for problems. No, make that we are looking for causes of problems."

They drew blanks with this approach too.

Day 35. Reply from NSEA. The reply finally came. It appeared that the NSEA engineers and scientists also drew blanks. They did provide a list of specific items that we should check. They also provided this directive:

1. The SC12 crew is instructed to maintain acceleration and speed at lower levels to avoid buffeting.
2. If the corrective action is successful, then SC12 is to continue on the mission as originally planned.
3. If the corrective action is not successful, then the mission is canceled, and Captain Logan is hereby directed to return for landing at NSEA campus.
4. When you have a final decision, you are also directed to immediately inform NSEA of that decision.

Captain Logan and his three helpers carefully investigated the short list of items mentioned in the first part of NSEA's message. That only took an hour. Nick said he wanted to tell everyone what happened and what they were going to do. "But first I need a new cup of coffee. Or a beer!" His comment created a small amount of levity to the otherwise serious mood. When he returned with his coffee, he said, "Bert, you can start with an intro."

Bert replied, "Will do," then keyed the mike and started, "Ladies and gentlemen, Captain Logan has some important information to pass on, so please give him your full attention. Captain Logan."

"Thank you, Bert. Ladies and gentlemen, yes, I have some bad news to share with you. You have undoubtedly wondered what went wrong to cause the noisy shaking and buffeting that we went through about two weeks ago. We have had some very smart people searching for the cause. No help. I sent a message to our leaders back at NSEA, expecting of course they would provide us with some corrective action. That was no help. As a result, we are canceling the mission, and will soon turn around and head for home. I have not yet worked out a new schedule, but my first estimate would be about a month. My first task is to turn SC12 around and head for home. Then I will prepare a new schedule and share that with everyone. That is all for now."

Day 36. Preparing to Turn Around. Captain Logan was not looking forward to turning around. This is a very complicated maneuver. Fortunately, the computers do all the work, so Nick and Bert would monitor the process in case something goes wrong. There is no atmosphere out here, so we can't just use a rudder to start a turn. We are traveling at L0.5, and we start a slow spin using the side thrusters while we are moving forward at one-half the speed of light. When we are pointed in the other direction, we will turn on the main thrusters.

But the first step is to tell the computer what we want to do. Nick pushed the button that said, "New Destination." He heard a buzz that was different from normal. Nick assumed the computer was acknowledging this important change of destination. He then pushed the other buttons to tell the computer to take us to Planet Earth for a landing at NSEA Campus. Then there was another buzz, and the AVC had a message, "INSUFFICIENT INFORMATION. UNABLE TO EXECUTE."

Bert had already located the appropriate manuals so they could monitor the turnaround process. Nick said, "Bert, I need you to look at this," Nick pushed the cancel button and the message on the AVC went away. Nick then told Bert to take notes while he repeats the process. Bert jotted down his notes covering each step. Then Nick said to Bert, "Let's take a look at the manuals."

Bert found the manual with the detailed instructions. They read through the instructions, and then tried for the third time, carefully following every detail of every step. The results were the same. Nick pressed "cancel" and they both watched. Finally, Nick said, "This makes two problems; or are we witnessing the results of a single, larger problem?"

"Well, whatever it is, we need to get a report to NSEA. And pray to God they have a solution for this." The two of them prepared a detailed report. Bert's notes were very helpful.

Day 36. Zero G. The turnaround was canceled. They sent a message to NSEA, and a reply would not be expected for at least another eight days. Nick and Bert went on the air to bring everyone up to date. Nick gave a short version of the problem, trying not to scare anyone. Bert said we would be at zero G most of the time. Brief periods of one G would be scheduled every two or three days, just to break up the routine.

"Let's have another 3-D basketball tournament!" The basketball committee put out a schedule for one-hour practice sessions this day and the next. On the following day they would have a new schedule for practice games, and then five or six days later they would have a schedule for the tournament. The winner of the tournament we held a month ago was awarded a plaque titled, "Champion of SC12." The winner of this new tournament would become, "Champion of the Universe." We only had room for thirty spectators at the finals. Another seventy or eighty managed to pack into the other court and the remaining meeting room. Those left out could still watch on their own AVCs.

The basketball tournament was exciting to watch. Several of the players looked good enough to try out for a team in the professional league - after we return home, of course. The day after the championship game was a boring day. Bert sent a message to the team leaders suggesting they look for other games they can play in zero G. Football? Soccer? Tennis?

Day 47. Suggestions from NSEA. Their report back to us indicated, “nothing definite found,” but it included a long list of things we should check into. Bert made four copies of the list and Nick called Hank and Betty to join for the review. Many of the items on the list were immediately identified as irrelevant and they were quickly crossed off. Some other items required going to places where they had not been before, and Bert called Lieutenant Bob Ventner, head of engineering and maintenance, to lead the way. Nothing was found there.

And then there was the computer section. Lieutenant Byron Williams is the ship’s computer expert, and Nick brought him in to help review sections of the computer code. Byron had an MS in computer science and had been on the NSEA staff now for ten years. The others had written some computer code in their earlier years, and some had been quite good at it. Bert bragged a bit about some of his past work, and said, “Once an expert, always an expert," but he soon had to retract that attitude, after he had asked, too many times, "What's that for?".

Hank said to the group, "I might have something here. In the code book, there is a section titled, Mission Changes. Go to page 29. Line 4006 is where it starts. Now, in the computer, select view mode, and then look at the same line of code. The code in the computer is somewhat different from that in the code book. Which is correct? Nick, you said the AVC said something to you when you tried to change the mission."

Nick replied, "Yes I just pushed the Change Destination button, and the AVC said *Insufficient Information, Unable to Execute*."

Betty jumped in, "Oh My God. Remember, I told you about the error I found in the manual. I reported the error, but it was never corrected. What if this is the new code that appears in the manual, but it was never entered into the ship's computer?"

Nick said, "This might be it. Byron, any comments on this?"

Byron answered, "There were some computer code changes made about a month before SC12 went up to dock at Columbia. I did not review the code itself, but I did look at the signoffs to make sure the right people were saying all required actions were completed and everything was as it should be.

Nick thought for a minute or two and told Bert and Byron to prepare a draft message to go back to NSEA about this, and "When the message is ready, let's reconvene this group for our final review and prepare to send the message."

After the message was on the way to NSEA, Byron started digging more into the computer code, "What has changed?" he wanted to know. In the code book he went back to the section for Mission, and then down to Change Mission. After that code there was a go-to function. That, in computer code, causes the computer to jump to a section titled Options and then the following was what appeared to be a list of selectable options. "Ah-ha, so the computer displays this list to the operator – that was Captain Logan – and he would select one of the displayed options." Byron then went to the computer so he could follow the code. After the line of code for Change Destination was another go-to function. Byron manually went to that location and the computer code said NOTE DO NOT USE-This section undergoing changes.

BINGO! Byron went to Nick and explained what he had discovered. Nick asked if it might be possible for Byron to make the changes. Byron shook his head, and answered, "Not possible. Changes are absolutely not allowed. I know for a fact there are traps throughout the computer code that watch for unauthorized changes and immediately remove everything that was changed. These were intended as safety features, so we don't have amateurs rendering our program unusable. I would also worry about messing up something so the computer is no longer operable. I would hate to be stuck out here forever."

Nick had Byron prepare a new report on the subject, describing the problem they just found. He sent the report as an addendum to the report he sent just a couple of hours previous. He also sent copies to Bert, Hank, and Betty.

What Next? Day 48. Two hours after the message was sent, Nick started thinking about what we should be doing for the next several weeks. He had planned on having joint meetings of the academic groups. For his flight to Pluto, he held these meetings during the cruise portion of the flight. That worked well, and he had planned for a similar schedule for this trip to Alpha-1. The computer code problem is preventing us from turning around to go back so we can have the correct code installed. And because of the buffeting problem, we are limited to 4.5 Gs, so it will be a slow trip, maybe six weeks back to NSEA if we can turn around. If we cannot turn around, what then? But Nick was not ready to bring up that subject.

"Eureka," Nick said quietly to himself, "Let's alternate between 3-D basketball and the meetings with the experts." He buzzed Bert's number and asked him to come in and help put together a schedule. They didn't want this to be too complicated, but they did think it important to make everyone as comfortable as possible. Bert had become close to those on the basketball teams, and in fact had asked Nick if he could join a team. Nick said he didn't mind Bert joining for the fun of it, but he didn't want him to be on a competitive team. "I would end up the bad guy if I had to pull you out of a game because I needed your help on something."

Bert showed Nick his sad face, and in his sincere voice he said, "I understand."

They would certainly not receive any complaints about the schedule. The mission was on hold, and they intended to keep everything as comfortable as possible.

Nick and Bert worked well together. Scheduled items were to be in the daytime, on alternate days. Except for Sunday, when the only scheduled items were church service. Basketball would start on the first day, meetings on the second, and so on.

Bert called a brief meeting of the team leaders. They were excited about the new schedule, and happy to participate. They all agreed it was too soon to schedule another tournament. The rough schedule, then, called for open courts on the first two basketball days, so to give everyone a chance to practice on their own. The basketball court was to be open at all times indicated on the final schedule, sometimes for individual practice and sometimes for team practice,

The court was to be closed from midnight until 7 AM. It was also to be closed during Nick's scheduled meetings, to encourage everyone to listen to or watch on their ACVs.

Work meetings were to be held in the meeting room. Each of our academic volunteers was asked to participate in at least one of these meetings. They were also invited to listen in on other meetings by their AVC. Comments and suggestions are welcome and are to be addressed to Captain Logan and Commander Castner. Nick said he wanted himself, Bert, Hank, Betty, Bob, and Byron to participate in all the meetings, and he planned to discuss with them what he hoped to accomplish in the meetings. The group of 200 academic volunteers would be divided into six working groups. Meetings would be scheduled starting on the second day, one in the morning, 8:30 to 11:30, and the other in the afternoon, 1:30 to 4:30. The purpose of the meetings is to discuss precisely what we should be looking for after we land on Alpha-1, and to identify any unexpected events that we might need to prepare for,

Meeting for Geology and Natural Resources. Day 51. The first working meeting started out just fine. The volunteers appreciated the opportunity to meet the Captain, in person, and up close. Captain Logan started with introductions. First was Bert Castner, the ship's Operations Officer, and Nick's first assistant. Then he introduced Hank and Betty

Hodge. and mentioned their past involvements in advancing space travel technology, He also introduced Lieutenant Bob Ventner, head of the Engineering and Maintenance Division, and Lieutenant Byron Williams, manager of the ship's computer experts.

Captain Logan then began his presentation, "As described in the earlier announcement, the purpose of these meetings is to discuss what we should be looking for when we are on Planet Alpha-1. We don't want to get side-tracked on some little things and then find out we were putting off some really important items."

"Alpha-1 does not exist on the Internet, so we can't ask Google about this and that. Google does not know these things. Our mission is to learn what we can. Is there life on Alpha-1? We don't know. Are there dinosaurs? We don't know. Will we have to do battle with raging lions? We don't know. Is the atmosphere breathable? Is there any oxygen? Is it too cold for us? Or too hot? Is it habitable? We have two hundred experts on the ship who know about these things, and we need them to take the lead in these meetings. Let's go down the list alphabetically. When I call your name, introduce yourself, talk about your field, and the important things you do. The rest of us will be asking questions. After a couple of minutes, we can move to the next expert. We have the meeting room for three hours, and that should be enough for everyone to talk openly."

Nick continued, "One of these important subjects has to be the composition of the planet. I believe that is geology. Andy, you have an MS in geology. What does a geologist do? What do you do?"

Andy Arnold replied, "I have been a geologist for twenty years. I started out working for an oil company in Texas, as a member of a small crew looking for likely places to dig for gas and oil. This is a science in itself, and it draws on everything we learned while earning that degree."

Another expert jumped in, "My name is Werner Gray, and I have been a geologist for only two years. My assignments have not been as exciting as Andy's. A typical day for me would be identifying the type of terrain and the types of rocks that are common in a particular area. This information is then useful to guys like Andy who may be looking for oil. Another important task for the geologists is to help draw a map of the world – as we plan to do on Alpha-1 – to show the land areas and water areas, mountain areas, forest or jungle, desert, and so on. If our cruiser is going to orbit the planet several times, we should have cameras running, so we geologists can prepare a global map and then we can start adding details to it. That should be helpful in many ways."

Werner continued, "Here's something we need to be very careful about. Let's say we are short on drinking water, and on the first day out we discover what appears to be a nice lake full of beautiful, clear water. We had better have the water tested before we all die of some poison we have never even heard of before. While we're at it, think about plants. Many of us are allergic to poison ivy. What if we have that kind of reaction to grass, and bushes, and other plants? Things like this can make for a miserable life."

Andy spoke up again, "Note that various areas may be listed as ecosystems, which may be categorized as mountainous, deserts, jungle, water, and so on. Natural resources are categorized in several ways, but I like the one I am most familiar with. This has seven categories:

1. Water is a source of energy, something to drink, and a place for water-creatures to live
2. Oil, coal and natural gas are major sources of energy.
3. Forests provide wood for building furniture and houses, energy, such as fire in the stove, as well as a place for other animals to live.
4. Rocks are great as building blocks; limestone can be ground up and used in making cement. or for grinding into several other useful materials

5. Minerals are useful for such materials as metal, glass, ceramics, and jewelry
6. Soils, for gardens and farms
7. Animals, food, wool, and leather

Nick was smiling. He was pleased with the way this was going.

Meeting to Discuss Fauna. Day 53. Nick's introduction started pretty much the same as the first meeting. Then he focused on the main subject for this meeting. "Here is something we need to be careful of. What if there is life on Alpha-1? If there are people, are they civilized or savage? Will they welcome us to their planet? Or should we return to our cruiser and make a quick exit? We could spend a couple of days orbiting and studying the flora and fauna. If there are animals, there is no reason to expect they would match the animals on Planet Earth. Back home, raccoons can be cute little critters. But what if the equivalent raccoons on Alpha-1 are larger and are ferocious meat-eating monsters? What if the place is literally crawling with man-eating snakes?"

Nick introduced his next expert, "Our first expert on animals is Tex Hanson. Tex is from Texas, and currently lives on a two-hundred-acre property, where he raises cattle and horses. Before he settled on the ranch, Tex spent twenty years as an animal trainer with several circuses."

"Hi. My name's Tex. I was born in upstate New York. My parents were circus people, and I grew up in the circus, fell in love with the circus, and eventually was assigned to work with the animals. I started out cleaning cages, and we all quickly found out I had a real knack for working with animals. The head man noticed I was pretty friendly with the new animals, and I was unofficially training some of them before the real trainer got involved. A year or two later, I worked as an assistant to the head animal trainer. Whenever the circus received new stock, they were mine for the first week or two."

"As you might expect, lions and tigers were the most difficult – partly because they have those sharp teeth and claws – and partly because they were used to having everything their way. And that's because everyone was afraid of them. I learned right away to never display fear. It is good to be cautious. No, all things considered, we must be cautious.'

"The first day with new cats was the hardest. My first task was to convince them I was the boss – and everything would be just fine as long as they did what I told them to do. I usually carried a sidearm for the first day, although I have never used it. You never know what's going on in a lion's head. He might have had a bad time with someone just before me, and I didn't want to be fighting someone else's battle. It's just me in the cage with that animal and he has no idea why I'm in the cage with him. It is up to me to calm him down, to make him feel comfortable; to make him accept me as the boss. At the right time, we would start training them to do what we wanted them to do.'"

"Enough of that. Let's talk about what's going to be out there when we step out of the cruiser and onto the surface of Alpha-1. You won't be training any animals – not in the first few days anyway. You want that lion to think you are just as bad as he is. Maybe worse. Don't give him any reason to believe he could survive a fight and be number one. You should carry a sidearm. Perhaps all the time. Others in your group should be carrying rifles, or shotguns.

"When we have our next meeting, I will provide you with some what-if notes. What if the lion just sits and roars? What if he charges? What if you are outnumbered?

There was a question from the audience, "Captain Logan introduced you as Tex. And you have a two-hundred-acre ranch. What happened to your circus career?"

Tex replied, "I was going to mention that. Thanks for the reminder. The circus business has been drying up. Circus owners are no longer in the mood to hand out promotions or pay increases. In fact, pay decreases are not uncommon.

"My uncle bought the ranch fifteen years ago, when it was much smaller. My folks quit the circus and joined my uncle last year. Then I did the same a month or two later. I toyed with the idea of training the cattle and horses, but learning my new job is taking priority right now."

A hand went up, "Another question, what was your nickname when you were with the circus?"

Tex answered, "I was known as Rodney all those years. I will answer to either name, but I prefer Tex."

Basketball. Day 53. The basketball players were delighted with the zero G so they could resume their new and favorite game, 3D Basketball. But there was an unexpected occurrence when we went to 1G for our first meeting. The basketball players still showed up, and they resumed the tried and tested game of regular basketball. Soon the basketball court was crowded with those practicing baskets from the circle. The team leaders contacted Bert and asked for his help to set up a new schedule so they could alternate between the two.

Bert first checked with the Captain to see how much leeway we had for this. Surprisingly, Nick said, "More zero G time means we save fuel. Granted, we are using nuclear fuel, so the difference is very small, but you remember the old saying, Waste not, Want not. Yes, let's go for more zero G time."

Earlier, when we had our basketball league competition, we had six teams. Now we were talking about two leagues, one named 3D and the other was called Regular. The two leagues would operate separately, different teams and different schedules. After we counted the number of those interested, we decided to have four 3D teams and four Regular teams. We required a minimum of seven players on each of the 3D teams, and twelve on the Regular teams. Both leagues played with five players on the court. The large difference in team size was due to the difference in physical exertion. Many of the 3D players were in for the entire game and never thought of sitting out for a few minutes. The Regular players were constantly on the move and most of them required some amount of down time. Another factor was age. Nearly all of our experts had been players in their college years, but many of them are now twenty or thirty years older.

The schedule was easy to work out. Captain Nick had originally established every other day as meeting days. Bert invited the two leagues to provide participants, and they started working on the schedule.

The new schedule for Mondays, Wednesdays, and Fridays called for zero G, and 3D basketball on the basketball court. The schedule for Tuesdays, Thursdays, and Saturdays was for our Meetings with the Experts, 1G, and Regular basketball in the basketball court. Sundays had zero G, church service in the morning, with the basketball court closed. Then the court opened at 1:00 in the afternoon for individual practice.

The 3D league was provided one week of team practice, one week of team practice games, and one week of tournament games, with the championship game in the early evening of the last Saturday in the series. The Regular league schedule was the same except for two weeks of team practice games vice one week.

Day 54. Nothing Heard. Captain Logan was anxious to hear from NSEA. He sent his urgent message eight days ago requesting help or advice on how to fix or bypass the portion of the computer software that was not working. He also sent updates every two or three days after that.

We were all anxious, hoping for a response soon. Nick sent a brief version of his urgent message to the whole ship. He was able to stay calm on this, "Be patient. I'm sure we will be receiving some instructions shortly." But Nick was running out of patience.

Nick placed a notice in the AVC system to cancel the next meeting with the experts, and instead called a meeting of those with backgrounds that may be useful for problem solution. Nick's list included education or experience in Aeronautical Engineering, Aerospace Engineering, Computer Science, Electrical Engineering, Mathematics, Mechanical Engineering,

Nuclear Engineering, or Physics. Nick said the purpose of this meeting was to familiarize everyone with the problem, and to try to come up with a way out of our dilemma. Others are welcome to attend if they feel they have something to contribute. Those not attending in person were invited to listen in on their AVC, and then following the meeting, anyone could ask questions or submit comments via the AVC.

During the meeting, they discussed the two problems. Nick had the floor for the first problem. "Initially we thought it felt like something loose and vibrating. But it was not speed related. When we throttled back, we were looking for a safe speed. We noted the buffeting stopped at 4.8 Gs. We then held the throttle just under 4.8 Gs, and we could slowly accelerate to higher speeds. So, the buffeting was not related to speed. It was related to Gs."

Back in the early days of our trip, Captain Logan had prepared his flight strategies. His goal was to accelerate to L4.0 (four times the speed of light) in the first month. To accomplish this the acceleration schedule started with eight hours at 2.5Gs for each day of the first week, and 3Gs for the second week. The third week would be scheduled for six hours at 4Gs followed by one hour at 8Gs. The fourth week, then, would be for six hours at 6Gs followed by two hours at 8Gs. The fifth week would include seven hours at 7Gs and one hour at 10Gs.

Initially, we were required to wear G-suits whenever we were above 2.5 Gs. Captain Logan explained that our tolerance will increase somewhat as we go along, and he would start testing us periodically. Now and then we increased, and we could soon handle 3 Gs without the G-suit. He said we would eventually be able to take 4 or 5 Gs for short periods. But we would need to be careful. Some show-offs ended up passing out at 4 Gs after a minute.

Eventually, we would reach our planned speed of L4.0, and then we could shut down the engines and relax. There is absolutely no friction in outer space, so we simply coast the rest of the way. Then, as we approach our destination, we will have to exert the same energy to slow us down as was required to reach cruising speed in the first place. We do that by flipping the cruiser around and using the engines to slow us down. If it took us a month to accelerate to our cruising speed, then it should take a month to decelerate. Nick said we will orbit Alpha-1 several times before landing.

"So, if you have been following this, you may still be wondering what the problem is. We have carefully checked the computer software and it appears the computer can take us through all of the required evolutions, including orbiting Alpha-1 and eventually landing on Alpha-1. The problem is, the one-way trip, which was going to take nine or ten months, will now take four hundred years."

There were forty or fifty people in the room, and most of those jumped up or held their hands up and started shouting questions or comments. Nick, calm, cool, and collected, held his arms out, "Ladies and gentlemen, please remember, the purpose of this meeting is to acquaint everyone with the problem. I just took you through the first problem. In a couple of minutes, we will acquaint you with the second problem.

"Now I would like to introduce Lieutenant Byron Williams, who is an expert on computers and is a computer programming expert as well. Byron is also in charge of the computers and computer programs on our ship. Please pay attention to the particulars of the problem, and then we can have a general discussion. Byron, it is your turn."

Byron took a deep breath, "Ladies and Gentlemen, I am Byron. Please bear with me on this. The basic system we have is just like those in the other NSEA ships. The operator tells the computer what he wants to do. Then the computer asks questions or shows a list of choices. The operator enters something, and the computer goes to the appropriate section in the program."

"We decided it was important to fix whatever was causing the vibrations, and we needed to cancel our mission and turn back to NSEA. Captain Logan was preparing to do that when we came face-to-face with problem #2."

"With his AVC in computer mode, Captain Logan selected MISSION. In a second or two, the computer responded with a list of options. The Captain selected NEW DESTINATION. The response from the AVC was INSUFFICIENT INFORMATION. UNABLE TO EXECUTE. There were several of us there, watching Captain Logan at the controls. We repeated the sequence several times, all with the same results.

A few days later, we found that the computer code depicted in the ship's manual differed from the code that was installed in the computer. At the point where the computer should be asking another question, the AVC showed, NOTE DO NOT USE- This section undergoing changes.

"It is apparent now that the updated code had not been installed in our computer. I was aware that some programming changes were being made, perhaps a month prior to sending SC12 up to dock at Columbia. The bookkeeping for these things is not easy, and I required two signatures for each code change. One signature was to certify the code was changed in accordance with instructions, and the other to certify the modified code had been entered where it belongs in the computer. Obviously, someone lied about completing his job. "

"The design of our system was set up so we could not deviate from the path to our destination without first making proper changes to the mission. And now I return this to Captain Logan."

"OK, I think everyone understands the problem. Here is the test question. What happens if we do nothing?"

Several hands were raised. Nick nodded toward the one closest to him, "What happens is SC12 shows up at Alpha-1 four hundred years from now with two hundred and forty-five skeletons on board." Several others laughed or chuckled.

Captain Logan replied, "Well, you are half right. SC12 will arrive there as you said, but the people on board will be alive. They will be our descendants, twenty generations down the line. Any questions?"

A look of surprise; and absolute silence.

The Captain continued, "I'd like to call it quits for today. But first let's talk about the schedule and homework. I would like to schedule our follow-on meetings to be every day, same time, six days a week. Your homework assignments are to come up with recommendations that may lead to a solution to our problems. Problem number one is the severe vibrations, or buffeting, that occurs at 4.8Gs. If we can solve that problem then we can be on Alpha-1 in a year, perform our mission, and back home in another year. Problem number two is the computer problem. If we can find a way to patch in the missing code, or possibly devise a way to by-pass that part of the code, then we can turn around, return to Planet Earth, and have proper repairs performed back at NSEA. If we strike out on both of these, then we will move our efforts to work on the four-hundred-year problem. But for now, let's all put our efforts to problems one and two."

Committee Number One - Vibrations. Lieutenant Bob Ventner had been with NSEA for ten years. He considered himself a mechanical engineer. In his late teens and early twenties, he had tinkered with automobiles and light airplanes. With NSEA, his management capabilities sprouted early-on, and he was usually head of the maintenance division or the power division. As such, he knows more about the mechanics of these cruisers than most. He called Ray Hilt to see if they could get together and work on solving this problem. Ray had earned his BS in aeronautical engineering and then was a pilot in the Navy for several years. Back in school, he earned an MS and taught for two years. Bob called Hank Hodge who jumped at the opportunity to help. Hank, in turn, asked Betty, but she said, "No, that's your cup of tea, not mine."

Bob was also able to sign up a few more experts for Problem #1. Red Jessen has an MS degree and was a well-known professor teaching physics. Dick Alderman has been teaching math and physics for ten years. Jerry Rumble earned his MS in nuclear engineering and distinguished himself with several patents for new designs. Paul Webber has an MS in electrical engineering and taught that subject for a number of years. Larry Bailey has a BS in electrical engineering and MS in physics. Mary Smith earned her MS in aero engineering and was anxious to help. Peggy Andrews has an MS with majors in both chemistry and biology, plus 15 years of teaching experience, and, most important, she feels she has a talent for problem-solving.

Bob was beside himself with excitement. He called for a brief meeting the next morning to share some of his ideas, and to invite all of them to attend the next "meeting of the experts" later the same day.

Then Bob called Nick to share his thoughts with him. Bob gave him a good overview. Nick's response: "Excellent!"

At the regular afternoon meeting, Nick opened the meeting by introducing Bob Ventner as the head of the Number One Committee. Bob then talked about his background and emphasized his intention to get the effort started and later, after a few meetings, they could elect a leader of their choice. He asked the other members of the committee to stand and give a brief of their qualifications and ideas for the committee.

Later that afternoon, Nick sent a message to announce the end of the meetings with the experts. His message also described the formation of two new committees. Committee Number One to be headed by Lieutenant Bob Ventner, and Committee Number Two by Lieutenant Byron Williams. Number One would have regular meetings on Mondays, Wednesdays, and Fridays. Committee Number Two would meet on Tuesdays, Thursdays, and Saturdays. Nick said he would probably attend some of the meetings. In any case, meetings would be available over the AVCs, and visitors would be welcome.

Bob had worked up a simple sketch of the cruiser, showing where the various pieces of machinery are located. Following a short discussion, Bob took the members of his committee for a walk so they could see first-hand what everything looked like.

Back in the meeting room, Bob told the group he wanted to tell a story. "My grandpa told me this story. At that time, he was driving a big Oldsmobile, one of the largest cars on the road. They had been visiting Iowa and were driving back to Virginia. It was early spring, and they encountered a snowstorm. They ended up driving in a foot of snow halfway across Tennessee. The big Oldsmobile was doing fine until Grandpa felt the car give an occasional jerk. He said it felt as if someone was stomping on the brake - but for only an instant, and only at the right front wheel. After several of these, he stopped. Everything looked as it should except the fender well was stuffed with snow. He cleaned out the snow from under the front fenders. Back on the road everything was fine for another five minutes, and then it happened again. This time he saw the fender well looked clean enough, so maybe that wasn't the cause."

"Back on the road, they saw a sign, Nashville, 15 miles. The wheel had developed a wobble, and Grandpa stopped to check things, expecting to see a flat tire. It was not flat, but the wheel was at an angle, as if there were no bearings. They drove on, slowly, and pulled in at a gas station. The mechanic was gone for the day. They left the car and carried their bags to a motel across the street. In the morning grandpa went to the gas station to see what had to be done. The owner said the wheel bearings were shot, and he showed grandpa a handful of small pieces that used to be bearings. They must have had a loose hub cap, and the snow sucked out the grease. Without grease, the bearings became extremely hot, melted part way, and tried to attach to each other, and then broke up again. The right front wheel spindle had to be replaced. The mechanic found what he needed at the local salvage yard. Using a cutting torch, he carefully removed the damaged part from the Oldsmobile and welded in the replacement section."

"I tell you that story just to let you know that bearings can go bad, and they can have some strange reactions. Keep that in mind while we walk around the equipment on SC12."

Committee Number Two - Missing Code. Nick had no doubt that Byron Williams was the right person to be the head of this committee. Nobody knew the ins and outs of the cruiser's computer systems like Byron. Nick and Byron put their heads together to pick a good team to research and solve the computer problem. Betty Hodge, with a Doctorate in physics and several years of experience developing new technologies in support of space exploration, said she would be delighted to help. Dick Alderman, with a BS in math and MS in mechanical engineering phoned Nick and said he wanted to be on both committees. Paul Webber earned his MS in electrical engineering after several years as a senior electrician and had a knack for troubleshooting computer systems. Bruce Thomas has an MS in computer science and ten years of experience developing advanced computer systems. Jim Miller has a BSEE, and an MS in computer science, plus experience as the lead engineer for system design and development.

The problem was simple to define. A critical part of the software had been removed for updating and, apparently, was never replaced. We want to cancel the original mission and return to NSEA for repairs. Because the necessary software was missing, we were not able to change our destination. The operations manual has a complete depiction of the ship's software, including the updated code as it should appear in the computer.

The solution, however, is not so easy. "Traps" had been inserted here and there for the purpose of preventing unauthorized changes. Several possible approaches were mentioned by the team:

- Create a routine that will help us to bypass the traps.
- Prepare a new code that will function as a substitute for the code that is currently missing.

Byron assigned three or four volunteers to each of the new tasks. They agreed to research the problems and then give reports at the next working meeting, two days away.

We felt we were being buried with problems. First, there was an error in the operations instructions. Betty Hodges had reported the error prior to the launch, but the correction was not made.

There are several "speed periods" that we are becoming familiar with. We sort of expected we might have one acceleration period to take us from lift-off to cruising speed. But things are too complicated for that. Accelerating at One G is very comfortable, but that would take forever for us to reach Light Speed (L1), and our schedule calls for our cruise speed to be L8 or L9. Captain Logan is leading us through a program so we can withstand higher Gs, and we can reach an acceptable speed more quickly. We have been up to 5Gs several times, for brief periods, but we still have a long way to go.

When we have reached our intended cruise speed, we can shut down the thrust engines and simply coast. There are no particles in space, so we have no friction to slow us down. We are far enough away from other planets so we have no G forces from them, Remember Isaac Newton's first rule: an object in motion will continue at the same speed and in the same direction unless acted upon by a force. While we are coasting, we are at Zero G. No speed or direction change. Our operations experts will do the math and calculate when we need to start our deceleration.

Day 110, Preparing for a *Long Journey*. On day 110, Captain Logan called an "all hands" meeting. We squeezed fifty people into the meeting room, and others were instructed to listen to their AVCs. Captain Logan opened the meeting with a brief review, "You will remember on day 10 we had some problems while we attempted to increase our acceleration. We finally discovered an error in the operations manual, and Betty Hodge helped us figure out the correct procedure. Thank you, Betty."

Captain Logan also reminded the audience of the other two serious problems. One of those was the mysterious buffeting when we tried to increase our acceleration above 4.8 Gs, and the other was the ship's computer would not allow us to change our destination so we could return to NSEA. "When I learned that part of the computer code was missing, I started to think all this could be a setup. Perhaps someone did not want us to return. But I quickly realized I was just trying to pass the blame on to someone else. Most likely this is the result of an honest mistake, and we may not find a workable solution."

"Several weeks ago, we created two committees to do further research on the problems and hopefully find solutions, Altogether, we had an abundance of talent on our committees. These people have excellent qualifications, but, so far, solutions have not been found."

"The good news part of the meetings was that if we do nothing, our ship will continue our flight exactly as originally planned; that is, directly to Alpha-1, and then to a safe landing. The big drawback is that we will be flying at a much slower speed, and the one-way trip will take four hundred years."

"I have already talked to the kitchen people and the recycling people. They have convinced me that our systems can handle the four hundred year-long trip. There are a number of topics, however, that we will need to research and then make some decisions. For example: Who is in charge? How do we go about electing a new leader? Do we need an organization?"

Captain Logan has been in charge of this group ever since we left home, and we all felt he should probably remain in charge, at least until things settle down, and then we may need to elect a new leader. For a parallel, we discussed the management organization of a home-owners association. The developer typically establishes the first set of rules, and then eventually backs out when they feel the new organization is capable of managing themselves.

We presently have 225 personnel in the ship. 165 of these are men and 60 are women. We need to make plans about adding children to our group. Since we will be adding children to our population, should we increase the total number? That will need to be researched.

Then we will need arrangements for schooling. Many of our passengers are professional educators, so that should be an easy problem to solve. Beyond a basic education, we will probably want to continue their education and training into those professions that will be useful or necessary for carrying on the lifestyles of their parents, as well as special needs for Alpha-1. Normally, medicine and law might be very appealing, but we should not need a lot of doctors, or lawyers. Although some might believe it would be nice to have a lawyer to help us with some of these decisions.

One of those in the meeting mentioned that acceptance is very important. In his words,

"Acceptance can be the first step of a solution. Try as we may, we cannot undo our situation. We can only move forward, not back." Nick stood and prompted a round of applause. Then he thanked member for that piece of philosophy and suggested we should all keep that in our minds.

Planning for Babies. After the meeting, Captain Logan asked Doctor Jack (Doc) Brown, the ship's Medical Doctor to join him, "Doc, I believe it's time for us to put heads together and start making some plans. How about joining me in my bunker, so we can keep the meeting private for a while. Any time this evening is OK for me." They agreed this was a good time, and Doc accompanied Nick to his bunker.

Doctor Brown completed his internship three years ago and has been with NSEA ever since. "You know, this is not my long suit. I participated in about a dozen births during med school, then I actually delivered two babies during my residency, and then one more last year in an emergency."

Nick turned to look at Doctor Brown, "Tell me about the emergency."

"It was party time at the house of one of our friends. About fifteen people were there. The hostess was pregnant – very pregnant - and had suddenly excused herself. Then she seemed to disappear. About ten minutes later, I heard someone holler, we need Doctor Brown! I went into the bedroom. She had already dilated, and we could see about half of the baby's head. I asked for someone to bring some clean towels, someone to prepare a pot of hot water, and someone to call 911."

Nick's eyes were now larger, "And?"

Doctor Brown continued, "Fortunately, it was a very normal birth. Within just a couple of minutes the baby was out, the cord was cut, and his mommy was cuddling him in her arms. The EMTs arrived. I identified myself as the delivering doctor and stepped back so they could wash the baby with a good antiseptic and confirm that everything was performed properly. I felt like a hero."

Nick finally shifted the conversation to his subject. "Doc, about a year from now we will probably have a couple of dozen babies to bring into the world, Will you and your assistant, Charlie Carter, be able to handle this OK?"

Doc replied, "No problem. By the way, Charlie was a senior OR Nurse (or operating room nurse in the pediatric ward of a Navy hospital, and he has a lot of experience delivering babies. I can certainly use his help in our new pediatric ward. About one out of a hundred births might be beyond my personal experience, and I am sure his additional experience will be especially helpful in those cases."

Nick was starting to feel better about the baby situation. They talked about the size of their population. The current level was 225 total personnel. Nick said he was comfortable with that level and was thinking we should use that as our target for the long term. Doc was agreeable at first, but then pointed out that our first group of babies would all be over the limit. After some further discussion, they agreed they needed a temporary maximum for a couple of generations, and then come back to 225 for the permanent limit. They agreed on a temporary limit in the 250 to 275 range.

Medical facilities currently consisted of two converted bunkers, now referred to as patient room and medical office. They decided to add two more bunkers, one to serve primarily as a delivery room, and the other for recovery. They were undecided about opening a fifth room for pediatrics but decided to save that decision for later. In the meantime, a fifth room was reserved, and would, for now, be available for storage and other miscellaneous tasks.

It was sounding like Nick was about to call an end to this session, and Doctor Brown brought this up, “Nick, just how do you decide who will be having these babies?”

Nick let out a gasp of air, shook his head and said, “OMG, talk about getting the cart before the horse. I’m certainly glad you brought this up.”

Doctor Brown replied, “Nick, you are good at communications. I can offer some suggestions, and then I believe you should take it from there. How about a message to everyone to describe what we are proposing? Then we could have a series of specific questions." Together, they worked up the following:

- The first section would describe what we are proposing, and a space where everyone can reply back with their comments regarding the "baby program."
- The second section would be a direct question to the females. Are you interested in having a baby? Possible answers might be, yes, no, undecided, maybe later.
- A third section with a related question. Have you decided on a man to join you for a baby? Possible answers could be the same as above, yes, no, undecided, maybe later.
- And then a section for participating man and woman: Shall we plan on a wedding? If yes, Catholic or Protestant?

Nick thanked Doctor Brown for his help and told him he would be included whenever more help was needed on this subject. Later that evening, Nick prepared his message and sent it to Doctor Brown for review. Doc agreed to everything in general, and then offered several other suggestions as well. Nick was pleased with Doc's comments, incorporated several, then forwarded the message to the other NSEA officers and his major helpers Bert Castner, Bob Ventner, Byron Williams, Charlie Carter, and Doctor

Brown – and also to Hank and Betty Hodge. He received comments from all; Most were brief comments: sounds good, right on, and so on.

In his final message, he made it clear he wanted to hear from everyone. “And “lease, send your replies not later than a week from now.”

Chapter 2 – Preparing for a Long Journey

Board of Directors. Nick had thought more about establishing an organization that could take over decision-making and management issues. We have a basic set of rules provided to us by NSEA. But those are intended for different circumstances, and they seem to follow the style of military organizations. Something along the line of an HOA organization made more sense to him.

The board will have to deal with more of these decisions at the next meeting. It is not that Nick wanted out of the loop for making those decisions. He just felt it was the right thing to do. He prepared a spreadsheet showing a cross-section of the present population, and then some additional numbers for years 2, 3, and 4, for the new babies, as well as some minuses starting in year 20 for the normal old-age attritions. At a glance this looked reasonable, but Nick would be the first to say he had no real basis for this.

Nick asked the three other operations officers to join him for a short meeting. He shared with them his thoughts about controlling the size of our population. They had no problem with understanding his ideas to increase the size of our population for a few years, and then bringing the size back to a lower level.

Nick then said he wanted to have a discussion for the Board of Directors. "We need to have people working on these issues who have some education or experience in those areas. One of the duties of the Board would be to establish committees of people who have qualifications in the subjects we would be addressing from time to time." This was a brief meeting, and Nick thanked the three for joining him on short notice.

Nick prepared a message to all hands, describing a method for electing the members of our board of directors. Nick said he would volunteer to be the initial of the board of directors. As the captain of the ship, he was currently the logical person to take on that responsibility. After some period, yet to be determined by the new board, they would call for a vote by the board members to elect a new president. Nick's message asked for comments and additional ideas on the board of directors. Hopefully, he would have sufficient responses to establish a board of directors in time to select a committee for the baby subject.

There were other subjects that came to Nick's mind. One that he felt needed to be eventually looked at was genetic problems. Our current population is only 225, and we do not know if that is large enough to avoid those problems. We have a number of people who are experts in computer programming and computer system design and development, but we have not yet found a way around the ship's computer problem. Should we have a committee on that subject? If we are not smart enough to figure it out, then maybe we can look for a way to become smarter. If we can find a way to make time travel possible, perhaps we could travel back and fix the computer problem before we start the mission. Nick's head was swimming with thoughts and ideas. He was feeling overwhelmed. We need a board of directors – we need a group to tackle these problems. We may need a different group to attack each individual problem.

For the present, Nick decided to keep these thoughts to himself, and perhaps bring these up after we have a board of directors. Then quietly, to himself, "And there will probably be other problems that I haven't even thought of."

Two days later Nick had received enough supporting comments for him to finish preparing his "plans" to establish the board of directors. Statement #1 discussed the purpose of the board. Statement #2 regarded Nick's volunteering to be the first president. Question #1 asked if the board should consist of 5, 7 or 9 members. Question #2 asked if the member wished to be a candidate for a board position. Next was a section where the members could nominate one or two members for the board. Nick asked for responses within the next three days.

It was apparent that the interest in Nick's "plans" had obviously increased. Small groups were gathering in the meeting rooms for discussions. Elevator traffic was the highest since SC12 had launched some four months ago.

Nick was anxious to start studying the baby situation, but the conservative side of him kept pointing out he was far from an expert on those subjects. It will be much better to pass that responsibility on to the soon-to-be created board of directors.

The early responses resulted in the following: The board will have 9 members. Nick was to be the first president. Nick thought it would be best if the board members were all volunteers, and he made a list of the volunteers. There were 41 volunteers. That was about one-fifth of the respondents. Then he went back through the ballots and counted the number of votes received by each of the volunteers. His goal, now, was to pick the top eight board members.

Nick was pleased with the votes. But there were some issues he still had to take care of. Several of the NSEA officers received enough votes to justify putting them on the Board. Apparently, they had impressed many of the passengers with their capabilities. Nick spent some time worrying about this, because it would not be fair if he started picking board members to suit his desires. Finally, he decided to consider them not eligible for board positions for the first few months. Nick was planning to withdraw from the position of president after two or three months, and he would put the question to the Board for deciding when NSEA staff members could be considered.

Nick did a search through the SC12 database for a simple version of Robert's Rules of Order and forwarded copies to the new board members. In his mail he also said he planned on three items for the agenda of our first meeting: welcome, ground rules, and committees.

Later in the day, Nick prepared his message to announce the new members of the board. The following was included in his announcement:

Capt. Nick Logan	MS Aeronautical Engineering, 10 years Navy, 15 years NSEA
Mary Smith MS	Aeronautical Engineering, 5 years teaching, 5 years industrial
Dick Alderman	BS Math, MS Mechanical Engineering, 10 years teaching
Hank Hodge	MS Physics, 4 years Army, 6 years space travel engineering
Andy Arnold	MS Geology, 20 years in the oil industry
Werner Gray	MS Geology 2 years industrial
Bruce Thomas	MS Computer Science, system development, 10 years of experience
Jim Miller	BS Electrical Engineering, MS Computer Science, 15 years
Peggy Andrews	MS, Chemistry & Biology, 15 years teaching

The first-ever meeting of the new Board of Directors was scheduled for the following afternoon.

Day 117, This Meeting Will Come to Order. The Board meeting was held in one of the meeting rooms. Board meetings were intended to be open to all, and anyone could attend in person, in the meeting room, or watch the meeting live from their own ACV.

Nick called the meeting to order and welcomed everyone attending as well as those listening and watching this, the first-ever board meeting of the SC12 Board of Directors. He then invited each board member to stand up and give a brief description of their education and experience.

“We have a lot to do.” Nick said, “We have some problems to solve, and some issues to discuss. I expect that all this will be much more than the board can handle by themselves, and I believe we need to establish committees to fully address individual problems and issues. The board will decide on a meeting schedule. Board members will attend each board meeting. Committees will be represented at each meeting and a committee member will give an updated status or progress report. Nick then said we needed a board decision for the schedule. There was a brief discussion, and a weekly meeting schedule was approved.

Nick continued, “And now we move to the subject of committees. You will remember about two months ago we created two committees. One of these was to research the vibrations we encountered as we were increasing beyond 5.1 Gs. We picked Bob Ventner to lead this committee and I will ask Bob to give a report at the next board meeting.”

"The purpose of the second committee was to find a way around the missing computer code that results in our inability to cancel our mission and return to NSEA. Byron Williams was picked to head this committee, and Byron will give us a report at our next meeting."

"We now have a need for another committee. I don't have a name for it, but it concerns preparing for babies, so let's call it the Baby Committee. Maybe someone will suggest a better name after our next meeting. I think we will need a lot of help for this. I am not aware of babies being delivered on a space cruiser. And I have certainly not imagined we would ever need to talk about living on a space cruiser for 400 years. All of you have seen the messages I sent out on this subject. I have another paper that summarizes the responses to those papers, and I will ask the board members to spend some time reviewing the summary. Then it will be the board's responsibility to start the process of establishing a committee to continue on that subject. I believe that may be the major subject for our next meeting."

"That is everything we had on the agenda for today. Can I have a motion for adjournment." The motion was made, the vote passed, and the meeting was adjourned. Nick let out a sigh of relief. He had worked hard to bring us to this point, and it showed. He was anxious to leave, go back to his bunker and get started on the next project. But first, he needed a nap. The members of the audience were crowded around, asking questions, and not allowing him to move.

Finally, he showed a nice smile and said, "I'm sorry, everyone. I was up most of the night preparing for this, and now I need a nap, I promise to be in better shape for our next meeting." Members of the crowd moved out of the way. Some shook his hand. Some patted him on the back. Someone started clapping and instantly they all joined in a nice applause. Nick turned back, smiled, moved on, and then disappeared into the elevator.

Vibrations Committee. Back around Day 60, Nick had decided to create their first committee. This was to try to chase down the cause of the vibrations and hopefully find a solution. If we could solve that problem our trip to Alpha-1 would take about seven or eight months, instead of 400 years. Nick appointed Lieutenant Bob Ventner as Chairman. Bob has a BS degree in mechanical engineering and is currently head of the maintenance and power divisions on our ship. Bob had recruited nine others, all with appropriate education and/or experience, to help in the research.

Nick called Bob and asked him to come by his office so he can bring him up to date. Bob was always very respectful when visiting his boss. At least, for the first few minutes. "Good morning, Captain."

Nick replied, “Good morning, Bob. Have a seat. I have fresh coffee here if you would like.”

“That’s a definite yes on the coffee. I watched the board meeting yesterday. I am hoping you are not thinking about canning the vibrations committee.”

“Not at all,” Nick replied. “I would like for you to have a committee meeting in the next several days, and then give a status report at the next board meeting. This does not need to be very detailed, but when finished, you should include your recommendations regarding what to do next.”

Bob answered, “Well, as you know, we have not found the source of the problem. But I think I will recommend continuing the committee, at least for another month or so. One of our first projects was to create a possible list. It is a rather long list, and we sorted the list into three categories: likely, possible, and not likely. We are about to finish with the second category, and anxious to start on the not likely list. Hank Hodge had an interesting comment at our last meeting: As the remaining list gets shorter, we need to consider changing not likely to very likely.”

Nick nodded his head, “I have known Hank for seven or eight years, ever since he and Betty were married, and he is undoubtedly one of the smartest people on the ship.”

That pretty much ended their discussion of the vibrations committee. They chatted for a few more minutes, and then Bob said, "Well I have some other things I should be doing. I'll see you at the board meeting, and thanks for the coffee, Nick."

Code Committee. About an hour later, Nick called Lieutenant Byron Williams, his appointed head of the missing code committee. Byron is responsible for the cruiser's computers, and all related systems, including the computer code. When Nick attempted to return to NSEA HQ to fix the vibrations problem, he found he could not change the ship's mission. The cause was traced to a section of the computer code that had been removed and should have been replaced with an updated version prior to departure on the mission to Alpha-1. Someone dropped the ball on that one. Byron signed up five other software engineers and developers to assist in finding a solution to the problem.

Nick was careful to explain to Byron exactly what he wanted done: have a committee meeting in the next few days; and, at the next board meeting, give a status report and recommendations.

Byron replied, "Nick, I will do what you asked me to do, but you should not expect me to recommend continuing with the project. I was part of the team developing the final version that was installed, and the system is full of traps to "catch" attempts to make unauthorized changes to the system, At the time I was very proud of the great job I did.

And now I wish there was a way for us to make those changes."

Nick answered, "Now Byron, don't you start feeling bad about that. You had a job to do, and you did it. Now you have another job to do. Keep looking for a way around the traps. Has Betty been helpful to you on this project?"

"Helpful? She has been the real lead for everything we've done."

"That doesn't surprise me. It is interesting that the two smartest people on the ship are married to each other."

Baby Committee. Nick was also pleased with the results from his questionnaire for the baby project. He spent several hours preparing a summary of the information returned to him from the earlier poll he sent to everyone about two weeks ago. He planned to send this summary to the board members, along with the agenda for the meeting.

- We currently have 225 people in SC12. We believe we can step up to 250 for the first two generations, and then drop back down to 225 for the remainder of the trip. The committee needs to review this and correct as necessary.
- We are thinking of 40 to 50 babies in generations 2 and 3. Is that a safe number? How often should we plan for more babies?

- In a few years we will be losing members from normal old age. Should we make any adjustments to the new baby goals?
- We currently have 60 females on the ship. 40 of these are capable of carrying a baby and they are all agreeable. Five of these are also agreeable to a second baby after a year or two.
- All of the agreeable females will ask for a wedding. Half Catholic and half Protestant.

Nick's instructions to the board members told them the Baby Committee would be responsible for the above action items. The board is not responsible for these. The board is responsible for establishing the Baby Committee and monitoring their actions. Names of the agreeable females will initially be treated as private information. A decision to release the names will be controlled by the board. Nick also included the plans he and Doctor Brown made for expanding the medical offices.

Fourth Month, Second Board Meeting. The meeting room was crammed; standing room only, with attendance estimated in the 80 to 85 range. Nick opened the meeting with a warm welcome. "Ladies and gentlemen, I am delighted to see we have so much interest in the important projects we are launching." Nick then spent several minutes talking about the relationship between the board and the committees.

"In the future - starting now – any member - any person on this ship – may make a recommendation or suggestion, to establish a new committee. Simply contact one of the board members, and we will put that on the agenda for the next meeting." At that board meeting, the board members would listen to any arguments for or against forming the committee, and then the board members would vote to start the committee, or not. The board members may ask if there is a volunteer to lead the new committee. If yes, the board may vote to accept that person, and then the new committee chair will present to the board a list of the proposed committee members. The board may then vote to accept or not.

"And now, the agenda. The first two items will be reports from the two committee chairs. The reports should include actions taken so far, actions planned for the near future, and a recommendation for continuing or ending. Following this discussion, the board will vote to continue or end the committee,

"The purpose of this particular action is to establish the relationship between the board and the committee. Specifically, it is the board's responsibility to establish and monitor the performance of any new committee. It is the committee's responsibility to research the issues related to the purpose of the committee and make recommendations for related actions."

Report from the Vibrations Committee. Bob Ventner gave the report. "I had expected this would be a straightforward and solvable project. Two pieces of metal that need to rub against each other need to be lubricated. If they run out of lubricant, they may break into pieces, or melt, or freeze where they are and jam up the machinery. I think we checked every piece of machinery on the ship, and we found nothing. When I talked with Nick just a few days ago, I was convinced there was nothing to be found, but Nick said something that turned me around.

"Nick asked me if Hank was contributing. I said that yes, Hank is good at coming up with new ideas. And then I remembered what Hank said. We looked at the likely sources of the problem and found nothing. Now we are ready to look at the unlikely sources. But we need to change our thinking. Those unlikely sources need to be re-categorized as very likely. And one more thing. We have examined all the places where two pieces of metal rob together. There may be additional places where pieces may unexpectedly make contact and then cause a problem. So, I recommend we keep working, and I promise I will pay better attention to what Hank says."

Nick thanked Bob for his report, and then asked the board members if they had any questions for Bob. They were silent, and Nick said, "In that case, we will count that as unanimous approval to continue."

Report on the Missing Code Committee. Byron Williams gave the report. "This is the most frustrating project I have ever worked on. As a ten-year NSEA employee, I was part of the team developing the software for this ship. One of the design requirements was to not allow changes to be made on a ship after it launches. We found a typed version of the missing code, and that is precisely what prevented us from entering that code into the SC12 computer. When we started on this project, we decided to focus on two possible ways to solve the problem. It didn't take long to decide neither of these would help. Unfortunately, we were discouraged and ready to throw in the towel. Now, we had another meeting just a few days ago, and we decided there are other possibilities to consider, and we are far from finished."

Byron sat down, and right away one of the board members asked, "What was it that convinced you to turn around?"

Byron stood back up and said, "Someone on the team said we are running out of ideas. We need somebody to come up with the ideas, and then we can research each one to determine if it can be a possible solution. And then Betty Hodge jumped up and said: That's where I come in. I may not be a programmer or systems designer but coming up with ideas is easy for me."

Byron told Betty she was welcome to attend their committee meetings any time she feels she has something to contribute, and then sat down. Nick smiled and asked, "Are there any more questions from the board members?" There were no questions.

Actions on the Baby project. Nick said, "That brings us to the final item on our agenda. The Baby subject. As you know I sent out a questionnaire. The responses were great. Almost everyone replied, and some of them offered some good ideas, which I will pass on to the Baby Committee. We have 60 ladies on the ship and 40 of those have volunteered to help bring us some new babies. I talked to Doc Brown, and he told me he and our Medic, Chief Carter, can handle this increase in population. Doc Brown and I also talked about expanding the medical facilities to include a delivery room and a recovery room. We have several empty bunkers so that is not a problem."

"Now, we need to create a committee to take over planning for these events. I talked with the board members earlier and our first step will be to ask for volunteers. So, let's see a show of hands from all who are interested. Then we can talk about who should be on the committee. For those who are watching on your AVC, press the join button, and you are live with our meeting." Five more volunteers checked in. Nick jotted down the names of those interested. Eleven altogether. "OK, now I'm going to call on you one at a time, and I want you to give a brief statement of why you want to be on the committee. Then we might ask you a question or two."

Nick called on Doctor Brown first, expecting of course that he would be volunteer number one. Doctor Brown said, "Well, I am not volunteering, but I do want to nominate the person who is best qualified to head up this committee, and that is my assistant, our medic, Charlie Carter. His experience includes 20 years as a Navy medic, with experience ranging from a battlefield medic for the Marine Corps, to a fully qualified operating room medic in a Navy hospital, and – most important – 5 years in the pediatric ward of another Navy hospital. After his Navy retirement, he has served as a NSEA medic, with another five years of experience on the NSEA space cruisers." Doc smiled at the group and sat down.

Nick nodded his head, and said, "Thank you, Doc. That is about the most impressive recommendation I have ever heard. Next on my list is Doris Patterson."

Doris stood up and thanked Nick for his hard work in bringing us to this point. Then she told her story, "My formal education is in aeronautical engineering. For some reason, it was an easy subject for me to learn, and I have gained a good reputation for teaching the subject as well as a practicing engineer for several manufacturers. But my first love is children. I have raised two children, now aged 14 and 20. I would have had more, but my doctor told me no more. When I learned I could no longer have children, it almost destroyed me. I started volunteering to baby-sit for some of my contemporaries with young children, and I enjoyed doing that. One of my friends was a certified midwife and she asked me to give her a hand several times when her regular assistant was not available. That became my new love, and I took a couple of courses, hoping to eventually earn my own certification. I have been a successful engineer, and I was hoping to become a successful midwife. Please, let me join the new Baby Committee."

Nick said, "Doris, thank you for telling us about your interesting background. I do believe you will be on the Baby Committee. Now, I saw a hand go up from Tex Harrison. I believe everyone knows of Tex's background as an animal trainer. Tex, we are all anxious to hear from you."

Tex stood, smiled, and said, "I have been very close to the animal kingdom for my whole life. I started training animals when I was 8. I was the lead trainer for new stock when I was 12. That includes some of the meanest animals I have ever met. I just had a knack for that sort of thing. Maybe it was easier because I also passed out their food at mealtime. I was the nursemaid when they were sick, and I was their midwife when the time came for little ones. Also – now catch this – I have also delivered four babies to our female employees. These were never planned for. These unplanned emergencies usually happened when we were traveling on our next visit, and too far from civilization to look for a doctor or a hospital. The circus is a small community, and word travels fast when it is time for a new tiger or lion to be delivered. I usually had a small crowd when the time came, so they called on me when one of our ladies needed help with a delivery. I am not volunteering to join the Baby Committee so I can deliver babies. No, I am volunteering because I understand the process, and I am good at figuring out what to do when the unexpected happens."

Tex bowed and then sat down. Nick commented, "Tex, I have only seen the rough and tumble side of you, and I never expected there was a sensitive side as well. Well done, and I am sure we will have a place on the committee for you."

In the next half hour, we heard from the remainder of the volunteers. Two had dropped out, so we now had nine left for the committee. Nick was anxious about the next step, selecting a chairperson. Then he spoke directly to the board members, and said, "Let's take a few minutes for discussion, and then we will call for a vote.

The discussion did not take long, and in a few minutes, there was a motion to elect Chief Carter for the position. The motion passed, and Nick congratulated the new committee chair.

Nick then described that he would be forwarding his notes on the subject to the chair, with instructions to distribute to the committee members. "You will see a few things in the notes that are to be considered private and confidential. People's names, for example, are to remain confidential until the board tells you different."

Nick thanked everyone for participating in this important meeting, announced the next meeting would be next week, same time and place, and the meeting was adjourned.

The next week was very busy for all three committees. Both of the two meeting rooms were booked for committee meetings twice every day.

Vibrations Committee. Bob Ventner was a good choice to lead the committee. He knew about everything there was to know related to the mechanical equipment in the cruiser. Hank Hodges and Peggy Andrews were great at coming up with new ideas.

In one of the earlier meetings, Hank said, "Here's an idea that came to me last night when I was trying to go to sleep. When we are accelerating, the forces try to move stuff: That is, loose stuff; stuff on the floor that may not be properly fastened. Some of this loose stuff could drift back and into some machinery that has a moving part. Then, when the Captain hears the noise and starts to decelerate, the G forces may push the stuff back to its original place. Later on, I fell asleep and had a bad dream, a nightmare actually. Something like I just described happened. It was in a room that we never go in. A box full of metal parts had been sliding back and forth, making a terrible noise while it was in contact. Then, after maybe the fiftieth time we ignored the loud noise and, instead, we kept accelerating. The noise got louder and louder, and finally there was an explosion. We were all killed of course. That was the end of the nightmare, but please don't ask me to describe the details. It gives me funny feelings when I think about it. I am really anxious to fix the problem. Let's put this at the top of the list.

Bob commented, "That's a super idea. I never thought about stuff sliding around on the floor. Let's plan a walking tour this afternoon. Any other ideas?"

Peggy held up a hand and said, "Mine is not nearly as graphic as Hank's, but it might be just as important. We have a lot of rotating machinery. Some of it is fixed drive, with steel gears used to transfer the torque from the driving piece to the following piece. Is it possible that something might allow a piece to move slightly, so the teeth in a gear could slip – maybe every other gear tooth. I am glad that Captain Logan decided to immediately start decelerating. You know, now I'm starting to feel like Hank described. My stomach feels a little uneasy right now."

Peggy started to sit down, "There is a similar problem that could happen where we have belt-drive equipment, but I can't talk about that right now. Maybe this afternoon."

Bob took over. "Hey gang, this is really great. I'm sorry about the sick stomachs, but both of these are certainly worth pursuing. I will prepare a sketch for both of these new ideas. Let's meet back here at two o'clock this afternoon. I'll also bring a pocket full of barf bags."

These two items, one from Hank and one from Peggy, gave them plenty to do for the next week. Just about every day, others on the committee would have another suggestion or two, but after only a few minutes of discussion, they were quickly dismissed.

Missing Code Committee. Byron Williams thanked everyone for showing up for the meeting. He turned to Betty Hodge, and said, "Betty, you said you are good with thinking up ideas. Now is your chance to save us."

Betty replied, "I have already started jotting down words - words that are related to the subject at hand. Then I went back down the list, using those words as titles, or headings, and added more thoughts to expand on the original.

"Here is one I particularly liked. I called it Fool the Computer. There are many programming languages. Which language is used for our program?

Does our computer understand other languages? Could we write some already sections that the "trap" would not know was there? And here's another possibility. Remember when Byron first went to the location in the computer code where our new destination would be entered, and what he saw was: "NOTE DO NOT USE-This section undergoing changes". Can we change the "go-to" coding so the computer will do what we want it to do?" Betty turned to look at Byron and gave him her "your turn" look.

Byron paused for a moment, then answered, "No, we already spent a good deal of time chasing that one. The problem is the computer will interpret this as modifying the program and delete everything we inserted."

Betty leaned forward and said, "Two up, two down. Not too good. Here is my last idea for today. Can we retype and replace the whole program? I know that takes a lot of time, but we have more time than we know what to do with."

"Betty, sorry, but that makes three outs. There is a password-protected area that contains a complete copy of the entire operations program. When someone tries to change the program, the changes are deleted, and then replaced with correct sections from the master copy. If we delete the entire program, a copy of the master program immediately replaces our new copy. I am very familiar with these parts of the program because I was the lead developer. I used to be proud of what I prepared for SC12 but now it makes me feel sick to think about it. Ugh, now I have a tummy problem."

Several other committee members presented more ideas, but Byron carefully described why their contributions would not work and thanked them for trying.

All done for the day, Byron looked tired. "Team, we need to dig harder. Let's plan to meet daily for the rest of the week. Same time, same place.".

That evening, Byron's phone buzzed. He glanced at the screen and saw it was Betty. He pushed the talk button, "Hi Betty, what's up?"

Betty responded, "I have an idea that I think might work. The way you described the safety procedures, you always said the program would sense something being changed and then whatever was added would then be deleted."

Byron said, "Yes, that is true. So, what is your idea?"

"OK, in order to change our destination, the first step is to push the button that says, New Destination. The problem is the go-to is supposed to take us to the New Destination section. But that section has been removed, so the computer posts the message, Insufficient Information. Unable to Execute."

Byron said, "You are right on so far."

"OK, the first step is to create a new button that says, for example, Change Destination, which should be just before the existing statement, New Destination. Shall I continue?

Byron said, "Yes, keep on going."

"OK, the next line in the program is a new go-to section. We shouldn't need a list of alternatives to pick from, so here are the instructions necessary to tell the computer to take us back to NSEA. The point is, we would be adding or inserting something and not changing or deleting anything. What do you think of this approach? Will this work?"

Byron replied, with a serious tone to his voice, "Betty, I don't remember any of our safety procedures looking for additions. I believe everything I was involved in was designed to react to deletions or changes. At tomorrow's meeting, I'll give a brief intro and then you can tell the group pretty much the same as you gave me tonight. There is no need for additional details. Then we can discuss and work on the next steps."

The Missing Code Committee has been on recess for several weeks. This new idea gave them plenty to work on, all the way to the next board meeting.

Baby Committee. Charlie called his committee together the morning after the board meeting. "Good morning, team. We have a lot of work to do. As you know we have 40 women volunteering to have babies. I would like our very first action to be a private sit-down with these people. I believe a one-on-one visit will be the best. Here is a list of their names. Now I want each of you to go down the list and sign up for four or five each. You may by now know some of these people personally. That may be a good thing, or perhaps it is not. I don't know, but the decision is yours, so just try to do the right thing."

"Within a few days we will prepare a rough schedule. Perhaps something like 15 per year until we hit 40. Or, 20 the first year and 10 more in each of the second and third years. We might refer to our current population as the first generation, and the new group of babies (all 40) as the second generation. Some of our volunteers said they might be interested in having a second baby. I don't believe we need to make any decisions on that right now, but please remind me to bring that up at some later date."

Tim Wilson, one of the retired professors from Johns Hopkins University, raised his hand, "Has anyone thought about our average life span? If eighty has been our average, will changes to our lifestyle result in a change to our longevity? And, if so, should we expect our life span to be longer or shorter?"

Charlie was prepared for this one, "Thanks for bringing that up. I certainly don't know the answer to that question, but I have been thinking about a closely related subject. Assume for a minute that we will all live to be eighty. Then eventually one-fourth of the population will be in each of the following age groups: 0–20, 21-40, 41-60, and 61–80. We are currently way off of that, and our total population is almost completely in the 21-60 range, with 3 or 4 over 60. Now, think about 20 years from now. Most of our current population will be in the 41-80 range, we'll only have a few in the 21-40, and our 40 new babies will all be in the 0-20 group."

"Now, think about our actions another 20 years later, when we should be preparing for another round of babies. Our volunteers might then include a few ladies in the about-40 age and a few more almost 20. Math is definitely not my strong suit, so I believe someone else should take over this part. Tim, does your doctorate in physics include a lot of math? Would you like to take the lead on our math problems?"

Tim grinned from ear to ear, and said, "Didn't you know that all of the great mathematicians were also great physicists? I would be delighted to take the lead for our math needs. This afternoon, I will start outlining an approach to the math problems you have identified, and I'll bring that to tomorrow's meeting. And I promise it won't require calculus to follow the steps."

Charlie had kept his committee for nearly two hours, and he was anxious to bring this to a close,

"Here is a quick synopsis of what we worked on today.

"First, we want everyone to get started with the interviews. Do one today. Do at least one every day until you are finished. Remember, it takes two to have a baby, so you should tell your volunteer ladies they can bring their male partners. In any case, we will need information on the partners. Also, note that many of the ladies provided some of this information in their response to the earlier questionnaire.

"Second, we all need to put on our thinking caps about the birth schedule. Tim will help us out with a math model for figuring out what is good and what is not.

"The third category will be for the medical aspects. We have not talked much about this, but this category will be for things like medical equipment, exams, delivery, recovery, pediatrics, and so on. If you don't mind, I will take the lead on that one."

"That's all for now. We'll see you tomorrow, same time, same place."

Mood Changes. Nick wanted to bring up a new subject at the next Board meeting, but first he wanted Bert's opinion. "As we have discussed, the ship will take 400 years to reach our destination, and we really have no choice in the matter. If we do everything right, we will die during our trip, and our 20th generation grandchildren should live to see the ship safely land on Alpha-1, 400 years from now. Most of us have accepted our fate and are eager to work out the best possible solutions. Some others, however, are in a terrible mood. One crew member has already committed suicide, and I heard some others are talking about doing the same thing. A mass suicide was mentioned, but that was quickly rejected. A few arranged for brief sessions with Doctor Brown. He is neither a psychologist nor psychiatrist, but he was successful. Some others sought solace with their church leader. They did OK, too."

Nick and Bert decided to go back to the earlier practice of their nine o'clock "updates" on the pri-comm. That seemed to help, but we wanted to do more than that.

Nick called for an evening meeting to discuss. Invited to the meeting were Bert, Nick's number two for cruiser business, the board members, Doc Brown, Mr. and Mrs. Hodge, the Baby Committee, and the leaders of the two other committees.

"We all have different backgrounds," Nick said, "We start out with different genes that are inherited from our parents, and then we grow up in different environments. These various factors then cause us to respond differently to changes in life. Some of the reactions I have seen, or heard about border on mutiny, and that issue, believe me, needs to be handled immediately. We need to move quickly, but carefully, so we don't damage relationships."

Doc Brown held his hand up, Nick gave him the nod, and Doc started speaking, "I had several psychology courses in medical school, partly so we could calm down patients involved in an accident, or survivors as they learn a loved one won't be recovering from whatever they had. The first message we want to give is we understand. Remember, they are grieving, and we want to help them feel better. It is important for us to convince them they will feel better if they can accept the fact that we will not be leaving the ship. We need to convince them the great majority of us have already accepted these changes. We are working hard to make the best out of this, and it will be better all-around if they can join us.

Tex signaled and Nick nodded, "It is important to talk to them, calm them down. One of my most important actions at the beginning is delivering food to these animals. This is a good step toward convincing them we have their best interests in mind. When a lion turns on me, I know right away I didn't do a good job. Then I have to start over. Oh, another thing, it helps to have a method. Work this out ahead of time, prepare a plan, and then follow your plan. Thank you." Tex sat down.

Nick replied, "Tex, thank you for sharing that with us. And yes, I can see the similarities you talked about." Nick said, to the group, "I should have invited our two church leaders." Then he turned to Bert, "Bert, you have been working with those guys.

Call them up, tell them what we're doing, and we need their help." Bert stepped to the back of the room, punched in some numbers, mumbled something, then gave the group a thumbs up. Bert rejoined the group and added, "They should be here momentarily."

The two church leaders were there in less than five minutes. Nick gave them a brief summary of what we had been talking about. Then he said, "I am asking all the group leaders to help those who are having a tough time dealing with the situation. We would like to develop an approach that all of us can use."

One of the church leaders answered right away, "The two of us were just this morning working on a church story that we could use in Sunday's service. You will remember one of the briefings all of us attended, perhaps a year ago, before being accepted for the flight. One of the key topics was our mission, which was – and still is - to search for new minerals or other items that might be useful back home. We had to commit ourselves to the mission – *So Help Me God* - in spite of any unexpected events. In essence, we were dedicating our lives to the mission. Well, Nick, we feel we are still on that same mission. The difference is, it will take four hundred years to get there instead of one year."

Nick jumped up, "That's it. That's exactly what we need! Would you be OK with us using your message as the key part for all of us?"

Rodney turned to Dwight, who gave a thumbs up. Rod turned back to Nick and said, "Nick, we would be proud for you to use what we have prepared."

During the next several days, mission seemed to be the most popular word through the ship. We were not changing anything physically, just helping people find a way to accept our dilemma. We will never forget those we leave behind, but we must fully accept that the right thing now is to move forward and do the best we can with our situation.

Our approach was very straightforward, but definitely not in a way that could make these people feel forced. All of us understand what we have to deal with. We certainly cannot change our environment, but we can change most of those sad faces into smiling faces.

Months Five and Six. This was a busy period. The Vibrations Committee was still charged up for inspecting different pieces of machinery. They had identified twenty-two additional pieces of machinery that could possibly have caused the vibrations. The plan now is to put two observers on each item while Nick and Bert pushed the power up to 5.2 Gs. The original experience was that buffeting started when the ship reached 5.2 Gs, and then stopped when it was lowered to 5.0 Gs or 4.8 Gs. The committee recruited several more people which then allowed them to place two people at each piece of machinery, but they had to run the test two times so they could test all twenty-two pieces. This required a lot of coordination, so Nick and Bert lowered to one G to allow our inspectors to safely change positions and then repeat. The test was further complicated since everyone in the ship needed to be in their G-suits as soon as the ship reached 3.0 Gs. This required setting a schedule and advising everyone on board they were required to participate in the test. During the first test, on both runs, the ship started buffeting at 5.2 Gs and quit buffeting at 5.0 Gs. Unfortunately, none of the twenty-two items were observed to vibrate.

Bob Ventner called a meeting of his committee. They were all frustrated. Bob was looking for ideas. “One option might be to test every piece of moving machinery on the ship. I studied the schematic and counted over four hundred pieces of machinery with moving parts.”

Hank Hodge raised his hand. Bob gave him the head-nod, and Hank said, “Like it or not, I believe this is something we need to do. But our process is very disruptive to the ship’s daily schedule. We will need twenty days to do the test, but that doesn’t need to be twenty days straight. How about one test per week? Or one test per month? We can create a priority list, with the most likely pieces at the top of the list."

Peggy Andrews said, "Wait a minute, you have forgotten, we have a dozen or so ladies on the ship who are pregnant. Our G-suits are probably effective in the early stage of pregnancy, but at some point, we might be threatening the lives of unborn babies, and possibly pregnant mothers as well. Also, continuing this thought, I would expect a new-born would not be protected at all,"

Jerry Rumble spoke next, "I'm just a nuclear engineer, so this is sort of out of my league, and my vote may not be worth much. But if we are going in that direction, I would suggest the one-per-week schedule is what we should use. One-per-month just drags this out too long."

Bob summarized, "All right, I think we are on the right track. Let's do all of the above. You can help me create a prioritized list, and then I'll prepare a one-test-per-week schedule.

Bob presented the plan at the next board meeting. There were several questions from some of the board members, and eventually the board approved.

New Approach for the Missing Code Committee. This involved creating a new program that would mimic the steps in the current operations program. As an ordered step was completed, a green light turned on. If a step was not completed, a red light was activated. If they successfully completed the entire program, a complete test would give a series of green lights, and then a red light would be seen when it came to the step to change the mission, because of the missing code. They estimated it might take three or four weeks to complete this much of the program. The next step was to install the safety devices that had the purpose of denying any changes to the program. Byron Williams had been personally involved in actual updates to other operations programs. And that is largely why Nick had picked him to head this committee. Installing the safety devices would take a week or so, and testing might take a month.

At each of the next eight or nine board meetings, Byron gave a brief report on the status of their current tasks. After testing the safety devices was complete, we came to the hard part. The purpose now was to find a way around the safety devices, so they could insert into the program the missing program steps. One direction to go might be to find a way to deactivate the safety devices. Another method might be to “disguise” a change or somehow “sneak” in the change. In Byron’s first report on this part, he said, “This is mind-boggling.” He then told the group he would be happy to invite some additional volunteers to participate in a brain-storming session. To his surprise, there were several who asked to be included at their next committee. Volunteers for this next meeting included Hank Hodge, Tex Hanson, Peggy Andrews, and Jerry Rumble.

Baby Committee. The Baby Committee had established an approximate birth schedule, with twenty babies in year two and another ten in each of years three and four. All of the new moms and future moms decided to be married, so we had a good time with those. One of our members was very handy with tools and managed to make, from scratch, a ukulele. He said his next goal would be a full-size guitar. He used plastic, of course, since we have no convenient source for wood. Before long, we had a better-than-average vocal group, and they met frequently to try out different popular numbers and different styles. Many of us started showing up at scheduled practices, just to watch and listen. It became a regular entertainment.

Someone asked if they should try to start a Music Committee.

Someone answered and said, "Committees are for solving problems, and this is certainly not a problem."

Someone else offered, "Then let's start a Music Club."

A year later, the handy guitar technician was studying how to build a piano. Someone commented, "That will take a huge effort."

He quickly responded, "Well, it looks like luck is with us. I just happen to have a lot of time on my hands."

Month Seven. The members of the Vibrations Committee were well along on the final series of vibration tests. Only three tests to go. For today's test, Ray Hilt and Rodney Anderson were teamed to observe a small room, which had some metal machinery anchored to the floor. It was a busy-looking arrangement with pulleys, gears, and chain belts. That machinery certainly was not about to slide around on the floor, but could something else slide to the machinery?

There was a wall just three feet from the machinery. Actually, the wall contained a row of storage cabinets, with heavy duty doors. Ray was an inquisitive kind of guy, and was opening each of the doors, checking to see what was inside, and then closing each door, and double-checking to make sure the doors were properly fastened. One of the doors he opened was not fastened. Ray gave the inside a quick look, and then started to close the door. But before closing the door he gently swung it outward and noticed it could actually touch the machinery.

Ray motioned to Rodney and said, "Take a look at this. I wonder what would happen if the door was open when this machinery started up?"

Rodney pointed to the edge of the door, and replied, "This spot on the door has been damaged. It looks like it has tangled with that drive belt already."

"I was about to shut this door, but I think we should leave it open and see what happens. Do you agree?"

Rodney answered, "Yes, by all means. We are here to find problems, not hide them."

They left the door unlatched, so it was free to swing back and forth. It was a heavy door, and normally positioned itself resting against the wall. Perhaps the acceleration would cause the door to swing open. They both took pictures with their phones. Ray forwarded a picture to Bob Ventner and called him on the voice connection. Bob said, "I would like to join you, but Nick is about to start the acceleration, and I don't want to delay the schedule."

When the acceleration started, both Ray and Rodney were carefully strapped down so they wouldn't be tossed around the room. They tied their cameras to one hand and hoped they could have enough control to point the cameras and tap the button to take more pictures at the right time. When the acceleration passed through four Gs, they saw the door move slightly, and then around five Gs, it was very close to the machinery, which had all wheels and belts moving. Almost as if by magic, it started. Ray phoned Bob, and said, "Listen to this." When the door was first touched by the chain belt, the door swung violently to the closed position. The door, of course, bounced back and then started a continuing back-and-forth cycle and a terrible racket. This was a loud noise, of course, and the loud “bangs” were of an irregular fashion. Ray now said to Bob, “I think you can report to Nick that we found it. And tell him to hurry up backing off the Gs – this is killing my ears!”

Bob Ventner called an afternoon meeting of the Vibrations Committee. He also invited Nick to join the meeting. Nick replied, “Thank you. I will certainly be there.”

During the meeting, Bob’s first action item was to cancel the remaining tests. Everyone cheered that item; everyone except Nick. He pointed out that Bob had worked hard to convince the Board that the test was necessary. We should now prepare a description of what we saw in today’s test and then recommend to the board certain actions, one of which would be to cancel the remaining tests. Bob said, “Thank you. Why didn’t I think of that?”

Another action would be to inspect all of the spaces that contained moving machinery in similar layouts and see if there were any more cases where loose doors might be a problem. That action commenced later the same day and completed the day after. Two more locations were identified with similar layouts. Nick told Bob to contact the ship’s machinist, who took on the task of fabricating latches to keep the doors positively fastened when not in use.

Hank spoke up and said to Bob, "With that problem solved, it would seem that we could now accelerate to our planned high speed and arrive at our destination in less than a year."

Bob looked at Nick, and said, "Nick, can we do that?"

Nick answered, "Unfortunately, no. When we had the vibration problem, we were limited to five Gs. As of a few months ago, we now have babies and pregnant ladies to take care of, and that limits us to two Gs. I have an action item on my desk, to see if we can modify some of our G-suits to give protection to the pregnant ladies, and also for the babies, but I believe we are stuck with the four-hundred-year estimate."

Month Nine – PDL. Nick recalled a practice that was popular in some of the military organizations and was very helpful. Every military position, especially staff positions, has a purpose – or a mission. The Pass Down the Line Book, PDL for short, was started by someone who thought it would be helpful to write down the major tasks, purposes, missions, and then add helpful notes, modifications, or changes from time-to-time, so the PDL remains up to date. It was routine for an officer to ensure his or her PDL was up to date as they were prepared to move on to a new position, and possible promotion. When the incoming person arrives, they usually have a day or two, sometimes up to a week, for turnover. If there is a good PDL, the turnover may be quite brief, and the outgoing officer can hand his PDL to the incoming officer, saying something like, "Here, it is all in the book. Good luck with everything, and goodbye."

Earlier, when the "mission" subject came up, in months four and five, a bell had sounded in Nick's head, and he was reminded of the PDL. Nick made a note to spend some time on that subject – as soon as he had some time to spare. Now, in month nine, he pulled out his earlier notes, and started preparing to acquaint everyone with what he wanted them to do. Basically, everyone – including the full range, from as low as a machine technician, and all the way up to the captain of the ship (himself) – should prepare a PDL book.

The PDL will be useful to the originator as a place to keep useful information, such as tips on how to accomplish difficult tasks, and lists of who to contact for this or that. When the originators reach a mature age, say forty or fifty, and they are about as knowledgeable as they can be, then it is probably time to bring in a younger person to start learning the job. The younger person can be thought of as an assistant, or apprentice, and in a training status. The goal here is to have a fully qualified replacement ready to take over when the originator is ready to move on to a higher-level position; or retire from the current position.

Some general suggestions for creating the PDL are:

- Start with a description of the professional position you are fulfilling, or more accurately, the position you will fill when the ship lands and we are on Alpha-1.
- Describe what you will do when you step out of the cruiser. What are you looking for? What notes do you need to take? Be specific.
- As other thoughts occur, add them to the PDL. If you change your mind, or you have a better idea, include that.
- Review your PDL periodically. A year or two – or forty - from now you may have different ideas. Be sure to include those.
- Cruiser staff will focus their PDL for operation of the cruiser; both in flight and on the ground.
- Other personnel will focus their PDL on their primary experience and education, and how that is relevant to their mission.

Month Fourteen. Nick prepared this into a message format and took it to the next Board meeting. There were a few questions, and then it was approved. That evening, Nick sent his message to all concerned. He carefully pointed out that this applies to everyone on the ship. He said he would be happy to answer questions, and he would be available for an all-hands live conversation if appropriate.

In the past week, Doc Brown delivered eight babies. Everything appeared as normal as it could be. Normal? This is not normal. We are making history. Just imagine the headlines of our local newspaper: Wishful thinking. During the next two months we should have another dozen.

> **BELIEVE IT OR NOT!**
>
> **EIGHT BABIES DELIVERED IN A SPACE CRUISER**
>
> **MILLIONS OF MILES AWAY FROM HOME**

Then our total population should be two hundred and forty-four. The overall mood could not have been better. The Baby Committee established a records manager position to start keeping track of our new citizens, and who belongs to whom. They also created a wedding license and birth certificate, issued and signed by the Records Manager and also signed by the mothers and fathers.

Vibrations Committee. The committee had been very successful in tracking down the source of the vibrations and Bob Ventner and the ship's machinist took care of the loose door problem.

In the following board meeting, Bob gave his committee report, announcing that, "The purpose of the Vibrations Committee has now been completed, and I would suggest that the committee should be dissolved." After a brief discussion, and a hearty "thank you" to Bob and the members of the committee, Nick announced, "The Vibrations Committee is hereby dissolved."

G-Suit Committee. At a later Board meeting, Nick presented his ideas for a new committee to investigate two new types of G-suits. One of these would be for protection to the expectant mother and the unborn baby. The other would be to protect the new-born children. Early discussions with Doc and Charlie seemed to indicate that up to two Gs would be beneficial for both the expectant mothers and the new-born babies; maybe as much as three Gs, but they could not recommend that without more investigating. What Nick and Bob were asking for was the formation of a small committee to research and possibly design the new G-suits.

There was a brief discussion, and the new committee was approved.

Bob Ventner raised his hand and said, "I would like to volunteer to be the head of this committee."

Nick, as President of the Board, replied, "Thank you Bob. You certainly did a great job heading up the Vibrations Committee, and I am certain you can do as well in this new position.

Nick turned to the Board members, "You heard Bob Ventner's offer to head this new committee. Do we have a motion?"

Dick Alderman raised his hand, and said, "I had the privilege of working with Bob in the Vibrations Committee, and I move we accept his offer to head the new G-Suit committee."

Hank Hodge said, "I also worked with Bob in that same committee, and I second the motion.

Nick expedited the formalities, congratulated Bob on his new assignment, and asked Bob if he needed help.

Bob replied, "I have already started a list, including Doris Patterson, Mary Smith, Ray Hilt, and Tex Hanson. I spoke with each of these individuals earlier, and they have agreed to be on the committee. Will this be acceptable to the Board?"

Nick then addressed the Board members, "Any questions? Any comments?" There were none. Nick turned to Bob and said, "It's all approved. You did a good job, Bob. Now don't forget your committee report at the next board meeting."

Year Three and Missing Code Committee. At another Board meeting, Nick, as President, asked Byron for his committee report.

Byron responded, "I am really disappointed that I have no good news to report. We have completed all of the tests of the new versions of the operations computer program, but we were not able to find any way around the security devices. With these tests complete, we are out of ideas, and we are also extremely frustrated. .What I propose now is to drop back to one meeting per month. We will run these as brainstorming sessions until we hit on something that appears to have some promise, then we will move to weekly meetings. Any questions?"

Nick said, "May I suggest, you could send out a message to all hands, give them a brief summary of the problem, and ask them to do some brainstorming and send any new ideas to you."

Nick continued, "Ladies and gentlemen, unless someone has some new business, I believe it is time to adjourn. Thank you for attending. Our next meeting will be four weeks from today, same time, same place."

Year Four - Education Committee. The second-generation babies have all been delivered, and our population is now two hundred and sixty-four. The Board had earlier recognized the need for an education committee to determine our education requirements. Martha Walljasper was presently on the Baby Committee, and she agreed to take responsibility for identifying teachers for grades K-12. In year seven we will need a teacher for twenty students in grade K. Those twenty students would then advance annually and eventually graduate high school in year twenty. In year nine, we would repeat the process for another twenty students, and again the same numbers in year twenty-nine. These same requirements would be repeated at twenty-year intervals. Martha said she would need at least three teachers to start in year seven and three more to start in year nine. We should not assume that each teacher can teach all of the required grades from K to 12. To be safe we would probably need three or four times that number. As we eventually recognized, the education activities would eventually turn into a full-time job for a number of people. Martha had prepared a plan for identifying qualified teachers for these positions and establishing schedules that would guide us for our first fifteen years of education. After providing several impressive updates regarding her research, Martha Walljasper was formally appointed head of the Education Committee.

Collegiate requirements will be a bit more complicated. Our original population includes thirty-two individuals with PhDs, ninety-five with MA or MS degrees, Forty-five with BA or BS, and twenty-five with no degrees.

The members of the Education Committee had a long discussion regarding professional careers for our new citizens. We all want the new generations to be just as professional as the original members of SC12. The moms and dads will be encouraging their offspring to seek careers in the same fields that brought them to this mission. Sons and daughters would then be enrolling in education plans similar to those of their parents. They did recognize, however, there may be good reasons to make changes, and they decided to keep records of the education plans for everyone. This would not be a strict requirement, where a parent needed to request approval for a son or daughter to take a different education route. Instead, the committee would review the overall education plans to ensure the great majority of the plans were close enough. If they found that too many were deviating, they would talk to the parents.

We wanted to pay close attention to possible candidates for the most critical positions, and we all agreed our two medical positions should be handled first. Doc Brown is our only MD, and, in his early thirties, he is among the younger ones on the ship. Doc will reach age sixty in our year twenty-nine. Considering K-12 takes thirteen years and college through med school takes another eight, we will need about twenty-six years to produce our first new MD. How about re-educating one or two of our adults? It didn't take long to figure out the best way to go on this.

Our medic, Charlie Carter, was fifty at the beginning of this mission, so we would like a way to produce a qualified replacement in about ten years. Charlie and Mary Smith had recently joined in marriage and Mary is currently pregnant. Charlie said she has a good aptitude for the position, and he and Doc Brown could probably continue her education and take her all the way to becoming a fully qualified MD in about five years.

Charlie coordinated with the head of the education committee to prepare a list of courses for Mary. Yes, if they work very hard, Mary could be an MD in five years. But that education process would be a full-time job for Mary, and at least a half-time job for both Charlie and Doc. They could slow this down to a half-time schedule and still provide a nurse qualification for Mary in five years, and MD about five years later. Charlie and Mary started making plans for their expected offspring to pursue a medical career. Doc married a young lady a couple of years later, too late to have a baby in Generation Two. The two couples then worked out a long-term schedule wherein an offspring would become a qualified nurse a couple of years prior to the senior doctor reaching age sixty. They worked up a rough schedule for the next hundred years and agreed this was workable.

Year Five – Developing the PDL Process. Nick and Bert have been brainstorming the PDL process. Nick just turned sixty and has been preparing Bert to take over as Captain when he retires. Bert has already been well-trained but defining a process will make it easier for others to follow suit in planning for someone to take over their positions. Bert is twenty years younger than Nick, but he is a very bright young man. In the middle of his college years, he signed up with NSEA for an arrangement he could not turn down. NSEA paid his next three years of college, including the award of an MS in aeronautical engineering. In exchange, Bert was obligated to serve at least five years as an employee with NSEA. A career in the space exploration business was what he wished for.

His first ten years with NSEA were exciting and included qualifying in some of the fastest jet aircraft ever designed. That lasted for a year and was followed by rocket-ship training for orbits around Earth, and then super-rocket-ship flights to the moon. The super-rocket ships were limited in speed to one-thousandth the speed of light. A trip to Alpha-1, four light-years away, in a super-rocket ship of that vintage would take four thousand years. Each way. New technology eventually brought nuclear power into the picture, and about five years before this mission, the Alpha-1 flight, we finally achieved our first sustained flight at L1, the speed of light. More improvements came, and new space cruisers were designed for flight in the L5 range. We had originally hoped to achieve speeds of L10, and possibly L12, during the Alpha-1 flight.

We had been looking forward to a very interesting two-year excursion to Alpha-1 and back. But now, the best estimate is four-hundred years. One way.

Nick and Bert worked well together. Nick was a good thinker at the high level, and Bert could fill in details faster than Nick could think them through. Their purpose today is to put together some guidelines that others can follow. Bert had some good inputs, and Nick was quick to say, "That sounds good," or, "I like that."

Bert had been the note-taker for their brainstorming, and Nick said to Bert, "OK, now read back what we have so far."

Bert gathered his notes, and started, "All right, here we go. For those still in their college years, we prefer they pick a career field that matches with one or both of their parents. The education committee will collect that information and periodically provide statistics to the Board. If we have less than a 90 percent match, we might ask some to change their education plans. Our purpose is to ensure we have the right talents to perform our mission after we arrive on Alpha-1. Also, prior to the end of college, we will encourage the college students to seek out current professionals in the same field so they can start a relationship. Students on these tracks would start out in a training status. About halfway through their training they would be promoted to assistant. Then when the outgoing person retires, the new person takes the status held by his or her predecessor."

Bert said he needed to rearrange the next set of notes. Nick poured two new cups of coffee.

Bert sipped his coffee, said, "Thank you," and then continued, "We want to see a professional relationship formed by the time the outgoing person reaches age 40, and the incoming person is at least 20. The PDL process should be well underway when the outgoing person is 50, and the incoming person is at least 30. Retirement may occur any time after the senior person is 60. Staying active past age 60 may be allowed if the individual is still of sound mind and able body. We will strongly encourage all to step down, to retire, at the age of 70."

Nick asked Bert to type up what they had done so far and send it to him. He planned to smooth up a few parts and take this to the next board meeting.

Nick leaned forward and said, "Now let's talk about me. I am 60 years old, and you are 40. As far as I am concerned, you already know essentially everything I know. You have been my backup for everything, and you have met all of the training requirements. I am not in the mood to step down at this time, although I will probably be ready to say, It's all yours, when I hit 70, and you will be 50."

"And now, let's talk about your current position - as Operations Officer. You are now 40, and it will be another 25 years before the first set of youngsters will be graduating from college. Then you will be 65. I am sure you will agree that 25 is too young to be the ship's operations officer. But if you could hold the job for another 10 years, that same new graduate would then be 35, the same age as you were when we started this mission. That might be a viable solution, if – if you can continue to be of good health and sound mind. Now, think about this. We have two lieutenants on our staff who are good officers. How about starting to train both of them for your current position? They can then be your backup in the event your health does not hold up, and also have them train for each other's position. When they are approaching the mid-point of this training, we can promote them to the rank of Lieutenant Commander and the title of Assistant Operations Officer. Then, as we have the first college graduates, we can pick three more to start training."

Bert was anxious to reply, "Great idea. Would you allow me an hour to visit with Bob Ventner and Byron Williams in the morning? The three of us can do some additional brainstorming of our own."

Nick agreed.

Bert joined Nick the following morning in the operations office. In keeping with their daily routine, they looked at all the gauges to check pressures, temperatures, and RPMs. They both agreed that everything checked out as it should, and Bert made the standard entry in the ship's log, "9AM, gauges OK."

Nick turned to Bert, "How did it go with Bob and Byron last night?"

Bert showed a frown on his face, "They want my job."

"How so?"

"Just a joke. I hope. Well, here we go. Right now – this year - we start training Bob and Byron as backup for my position. I will also be training on the other systems to be backup for them. That is in case one of us doesn't make it to age 70. Now, in the year 15, you will be at age 70, you retire, and I step up for your position. That leaves Bob and Byron to handle the other three positions. That won't work so we need to bring one more person in for training. This year. One possible candidate for this position could be Hank Hodge. He certainly has the right aptitude, and he seems to be a quick learner. Another possible could be the guy from Keokuk, the wise guy. Ray Hilt. I think he is an OK guy once you get to know him. Anyway, I think he would be just fine as head of engineering. He is good with mechanical things."

"And here's something we didn't talk about. Who should be deciding which person is going to fill which position? I could make those decisions after I step up to be Captain. Or, better yet, you and I could decide prior to your retirement."

Bert was anxious to finish his summary." I have some more, but how do you feel about what I said so far?"

Nick looked very serious for a moment, then he smiled and replied, "You guys did a good job. I agree with everything so far."

Bert answered, "Well the guys were delighted to help with this. We worked hard, but we had a little fun with it too."

Bert was then back to his summary. "In year 15, I will be age 50 and just promoted to Captain. We promote one of the other three to Operations Officer, and the other two as heads of Engineering and Computers. Now, when we move to year 27, the year we have our first college graduates from the second generation. In year 27, we pick four or five recent college graduates to start training. We have 4 officer positions to fill, and I thought we should have one extra. In the year 35, I will turn 70, retire, and one of the others is promoted to Captain. In the next two or three years we will have to replace the remaining old-timers. In order to retire everyone by age 70, one or two of the new graduates may have to be elected to the officer positions around age 33."

Bert gave an exaggerated sigh of relief and said, "And that is the end of the hard part. In the later years, we should have plenty of college graduates to pick from, and we can even start encouraging people to retire at age 60."

Bert prepared a typed version of what they agreed to and sent it to Nick. Nick prepared a more simplified version to present to the Board. In that version, he made it clear that all officer candidates will be carefully selected and trained. And the PDL process will be followed.

Chapter 3 – Games

Basketball. A special meeting of the 3-D players was held for the purpose of imposing some special rules. The first change was unanimously agreed to. The agreed-upon change was simply to consider holding onto the basket, backboard, ceiling, and anything attached to the ceiling as out of bounds. The offending team would be charged with a penalty, and the other team would receive a free shot. Some clever 3-D players developed a new tactic where a player would take a floating position at or near the basket. Another player would throw him the ball and then the basket was given. Some of this seemed to be OK, but when the floating player takes a stationary position near the basket, he is blocking the other team from shooting a basket. We decided that will not be legal.

Both regular and 3-D Basketball continue to be great pastimes. We even have a few talented individuals who participate in both. Bert arranged to have these scheduled in alternate time periods. In off-season periods we still had a good number who liked to play for the fun of it, or for exercise. After a few problems with the BB players trying to play at the same time as scheduled meetings, we set up a standing schedule where the rooms were primarily for meetings in the mornings and until mid-afternoon, say 3PM. Then the rooms were primarily for BB for the remainder of the day.

We planned to hold four BB seasons each year. Each season would be made for four weeks, alternating between Regular BB and 3-D BB. Three weeks were provided for warmups and practice, and the fourth week for tournament games. Schedules were adjusted so the final championship game for each season was on a Saturday evening. No one seemed to tire of either playing or watching, so this routine may last for the entire flight. 400 years!

In our third year, someone found several dart boards and a few dozen new darts. According to Bert, these had been included with a load of spare parts, and then the darts and dart boards were forgotten. Until now. We placed three boards in one of the meeting rooms. We didn't even have to tell people they were there. As soon as one person saw the dart boards going up, we had people standing in line to throw darts. Darts became so popular we asked the talented guys in the tool room to make two more sets, wherein each set included three boards and three dozen darts.

Baseball. We had a few people on the ship who used to be avid baseball players. They would gather occasionally to throw the ball back and forth. They even had ball gloves. When asked, "Why did you bring a baseball and a ball glove?" The answer could have been expected, “I don’t go anywhere without my equipment.”

Just a few days later, we heard a loud “thump.” Then a minute or so later, a repeat. On stepping into the room, we saw the guy who was always bragging about his college days as a pitcher had painted a small bullseye on the wall, “I used to be a pretty good pitcher, and I don’t want to lose my skills. I am anxious for one of those zero-G days, so I can find out if that makes my pitch better or worse.”

A few months later, 3D-BB events were on the schedule for a while, and that meant zero G. Our “pitcher” showed up with ball and glove, “I can’t wait to try this out.” His first “pitch” was just a mild toss, to test the effect of no gravity. His first mild toss had no arch at all. The ball followed a straight line to the wall, and then straight back, passing several feet over his head.

The next pitch was a slow ball; and was carefully aimed at the center of the bullseye. There was a quiet “thump” and the ball bounced back to the pitcher. He caught the ball with his glove, without having to move left or right. He pitched a couple more balls, accelerating his speed each time. He noticed he now had an audience.

He turned to his audience, took a bow and said, “And now, let’s see what happens at ninety-five miles an hour.” He held his glove up to his face, bent forward slightly, wound up and threw the ball as hard as he could."

"Wham" went the ball against the wall, and then it returned, just as fast as when it left the pitcher's hand. He was so surprised, he froze. The ball hit him in the solar plexus, dead center and just below the rib cage. He didn't fall, he just sort of collapsed to his knees. A minute later he was back up, "The last time I was hit like that was a line drive from someone's bat. I think I'll sit out the rest of this inning."

Five minutes later he was back on the pitcher's mound, ready to throw some more pitches. Someone asked, "How's it going?"

"It still hurts, just a little, but it's getting better. No more fast pitches. I'm keeping this down to around seventy-five, or eighty. On second thought, maybe I should quit. This is not fun anymore."

He picked up his carry-bag, inserted his glove and ball, and closed the zipper. On his way to the elevator, he was heard to say, "Did anyone bring a football?"

Computer Games. Everyone on the ship was computer-savvy, but only a few were skilled at the popular computer games. All of us had one thing in common: too much time on our hands. Eventually, the novices were seeking advice from the experienced game-players. Before long, the novices became experts, and then we had a series of competitions.

After a year or two, several of the experienced computer programmers were developing new games.

Computer Golf. Someone asked about computer golf, "Hey, can you experts set up computer golf for us?" Computer golf originated nearly a hundred years ago, and many versions were available in the Internet. NSEA is a US Government organization, and they were able to make available to us detailed information on practically anything that existed on the Internet. That does not include ordering something to be delivered to us, although that was the first thought that occurred to most of us when we heard about it. But anything that could be downloaded to our AVCs was available.

Byron formed a small committee to work on this project. Bert asked, "There is a set of golf clubs in the spare parts room; can they be used?

Byron replied, "Certainly, or at least I think so. I know that some of the set-ups I've seen have the golfer hit a real ball with a real club, and then special electronic devices compute the distance and direction of flight. Done right, these were very realistic."

Downloading and setting up the software was the easy part. Building the place for the golfer to stand, and the large screens for images of the tee, sand traps, green, and other backgrounds were all interesting projects within themselves. Setting up the electronics was another fun project, although it became a bit tedious because we had to take a trial-and-error approach for making the very sensitive adjustments.

With everything set up and operating, Byron's team still had to spend several weeks on final adjustments. Finally, everything was operating as it should, and we had the use of a fairly easy 9-hole course. After a few months of very satisfactory operation, they started making improvements. The players then had the option to select from several 18-hole courses, some easy and others not-so-easy. We also had the option of running a tournament with up to 8 players. We were having so much fun, we started teaching some of those who had never played golf before.

Chapter 4 – School is Open

Years Six and Eight for Pre-School. Our first 20 children made up half of Generation II, now called GenIIA, and started the pre-school program in Year Six. Now averaging four years old, they were raised by their parents, who participated in informal events, so the children had ample opportunity to get to know others of their age. GenIIB students would then start in Year Eight.

We started the pre-school setup in the middle of the year, taking over one of the meeting rooms. Bert called in his construction team to start building another meeting room. They were happy to have something interesting to do.

As normal, parents were responsible for instructing their children in potty-training, mealtime table courtesy, and just generally getting along with each other. Pre-school teachers were responsible, of course, for teaching the alphabet and numbers. For the first half of the year, classes were scheduled for four hours a day, and then increased to six hours at mid-point.

Year Seven, Starting Kindergarten through Grade Six. We added two more teachers, one in year seven and one in year nine. Class days are seven hours. Kindergarten teachers are responsible for basic spelling, counting, and discipline.

Year Fourteen, Starting Middle and High School. We started with four qualified teachers, then quickly added four more for the high school years for special courses, foreign languages, and advanced math. Class days remained at seven hours.

Year Twenty – College Starts. During the senior year of high school, in year 19, the Education Committee started counseling sessions with the 12th-grade students. These meetings have several purposes. One is to encourage the students to follow the same or similar career fields as their parents. Another is to ensure that their grade averages are sufficient to support their plans. Yet another purpose is to seek out capable candidates for the two medical positions, as well as five candidates for ship operations.

By this time, some of the students had already arranged with someone to serve as their "big brother," or "big sister," and assist them in understanding the duties of the positions they wanted to pursue. The Education Committee had brought in Tim Wilson and Lee Stieber, both retired PHD-level professors with excellent teaching experience. Their primary tasks were to recruit professors for teaching positions. They were both now at age 75, but it was not an intense workload, and they seemed to do just fine. Two years later the process was to be repeated for the GenIIB. Both Tim and Lee declined to do a repeat, but instead gave instructions to others for these responsibilities, and that worked well.

Special Schedule for Medical Positions. Charlie was anxious to get started with Mary's classes that would provide her with another BS degree, this one with a Nursing certification. As earlier agreed, a somewhat leisurely schedule on a half-time basis would require five years. Following that, in another five years she could be an MD. Classes were scheduled for five hours a day, four days a week. Doc was the duty professor on Mondays and Wednesdays, and Charlie handled Tuesday and Thursday classes.

School started for Mary in year four. They followed the schedule, decided to skip the summer vacations, and she received her Nursing certificate in year 8. Mary decided she needed a two-week vacation so she could spend some time with their four-year old daughter.

Mary found the classes leading to the MD were definitely more demanding, and she was sacrificing some of her leisure time for reading assignments and homework preparations. Doc and Charlie were still alternating workdays. In her spare time, Mary took some of the load off of Charlie and Doc by performing some necessary nursing duties.

Mary graduated in year thirteen, and she was anxious to step in and work with Doc Brown. Their daughter was now nine years old, and Mary was looking forward to the year twenty-two, when she and Charlie wanted to plan on another child.

Doc Brown had married several years later than Charlie and Mary and were too late to plan for a child in Generation II, but their first child was one of the earliest in year 24, as part of Generation III.

Charlie Carter reached age 70 in the year 20. This was the age we considered mandatory retirement. It was time for Charlie to retire, although he still continued as a volunteer and helped when needed, until health would not allow. Their daughter was now 17, graduating from high school, and then planning to take biology and chemistry in a pre-med program. Two years later, their second daughter was born, and Charlie turned into a stay-at-home-Dad. This daughter became good friends with Doc Brown's son, born about the same time. Life is good.

Nick Logan will reach the age of 70 in another week. He and Bert had worked out their process for handling the new assignments and promotions for their staff. Today, then, is the day for reviewing and finalizing these changes, and this meeting was limited to Nick and Bert.

Nick started the meeting, "Well, Bert, you have worked hard to learn all there is to know about the Captain's position. I am ready to pass the job on to you. Are you ready to accept?"

Bert looked sad, "Nick, I should be happy to accept the job, but my emotions keep getting in the way." Bert paused for a half-minute, then continued, "My answer is yes. I am ready. Now you make sure you stay nearby, because I will probably want to check with you now and then before making important decisions."

Nick chuckled, "I'm sure you will be able to reach me anytime you need. But here is a piece of advice. Do not be afraid to turn to your assistants to help make difficult decisions. Let's move on, and work on your current position."

Bert now had his happy face on, "Ok, two of our current staff have been in line for this position ever since this mission started. Lieutenant Commander Bob Ventner is the current Mechanical Engineering Manager. He has been doing a great job of managing his division and the Vibrations Committee. Lieutenant Commander Byron Williams has been the Computer Science Manager for the past fifteen years, and the leader for the Lost Computer Code Committee. You know, I really felt sorry for Byron. That problem was created by someone not on the ship and there are no instructions. He handled that just about as good as anyone could. Ten years ago, we promoted these two guys to Lieutenant Commander, and told them to start learning the functions of the Operations Officer to be a backup for me, and also to take the Ops job when I stepped up for the captain's job."

Bert looked like he needed a break. He took a sip of coffee, breathed a deep breath, and then continued, "Nick, my vote is for Bob Ventner. Either one of these guys can probably do the job, but I think Bob has the broader shoulders to handle anything that might come up. How about giving Byron a lead position for training new people?"

Nick quickly said, "That's a great idea."

Bert turned to Nick and asked, "And who would you like to see in these jobs?"

Nick answered right away, "Oh I agree with your selections. Good job. Let's keep going."

Bert continued, "Well, that leaves two people. We brought Hank Hodge and Ray Hilt in for training positions twelve or thirteen years ago. Ten years later we promoted Hank to Assistant, with the rank of Lieutenant. I believe we should promote Hank to take Bob's job as Mechanical Manager, with the rank of Lieutenant Commander."

Bert turned to Nick, "Nick, do you agree with this?"

"Oh yes, I agree. Good job. Please continue."

Bert continued, "That takes us to Ray Hilt. Ray may not be as good as Hank, but he is certainly good enough to take one of the positions if someone must drop out. I believe it is time to promote Ray to Assistant, with the rank of Lieutenant. I plan to encourage him to continue training for both Manager positions. If all goes well, we can probably promote him to one of these positions in year 35 when I must step down as Captain. Nick, how about a short break. OK?"

Nick nodded his head, and then focused on his logbook.

Bert refilled his cup, then sat back down, stretched, and continued, "We will run out of Generation I candidates in year 40. To handle that, I propose bringing in four trainees in year 27. These will be from the first graduating class. They will be taking over in the year 40, so we will have to work hard to prepare them for those responsibilities. I would consider them in Training status for the operations, mechanical, and computer science positions, and maybe promote them to Assistant status perhaps in year 35 or 36."

Nick and Bert continued the meeting for another hour, to discuss the schedule for Nick's Birthday.

Nick's Birthday. A birthday announcement was distributed via the AVCs. The birthday party was scheduled to start at 10AM, live on the AVCs, and attendees were invited to the meeting room, up to attendance limits. The schedule included the following (times are approximate):

10:00 – Happy Birthday and Cake (by Bert Castner)

10:15 – Bert's Promotion to Captain (by Nick Logan)

10:30 – Nick's Retirement (by Bert Castner)

10:45 – Other changes (by Bert Castner)

The meeting room was stuffed to standing room only, with almost a hundred people present. Bert had the podium first to welcome the crowd to Nick's birthday. The cooks had prepared 20 birthday cakes, and they were readily visible on a side table in the room. Bert said he hoped that those watching on their AVCs could find a few minutes to stop by later to take a piece of cake back to their bunkers.

Nick took the podium for a few minutes so he could officially promote Bert and discuss his qualifications for the position. "My message is very clear; Bert is well-qualified for the position, and we can trust him to manage the ship, the staff that works for him, and all that is necessary."

Bert was at the podium again to say, in his formal voice, "Ladies and gentlemen, I want you to know I am honored and excited to take on the duties as Captain of this ship. I have been the Operations Officer for twenty years, working as the captain's number one Assistant. Please believe me when I say Nick has been very thorough with his training." And then, with a big smile, "In addition, Nick told me I could call him with questions at any time. And he also promised not to make any commitments that would take him away from us." Bert took a half-step away from the podium and nodded to indicate he was done – at least with this part. There was a good round of applause. Bert showed a smile, waved at the crowd, then he took a bite of his piece of cake.

Back at the podium, again, Bert waved for everyone to be quiet. With his happy face, Bert continued, "Yes, we have more to celebrate. We are celebrating not only Nick's birthday. We are also here to celebrate Nick's retirement. It is obvious that we all love our Captain Nick. We started this mission with Nick as Captain of the ship. But Nick has been handed problems that no ship Captain has ever imagined he would need to handle. Nick has demonstrated to us he is capable of handling anything. Nick has been the leader in defining the leadership and management functions for our community. This goes far beyond flying and operating the ship. It has to do with addressing some serious ship problems as well as everyday people problems. We established some rules so we could have reasonable control over our lives. Nick was the brains behind establishing the Board of Directors, and then forming specific committees to handle this and that. We have the Baby Committee. The Vibrations Committee did its job. The Missing Code Committee learned there was nothing we could do about that problem. We have an Education Committee. And who knows, we may need some other committees as well."

After a few more words about Nick's retirement, Bert acknowledged again, "Everyone loves our Captain, and we are all sorry to see him step down from his position. At the same time, we are happy to see that he is still in good physical shape so he can enjoy his retirement. And maybe we'll see him on the golf course."

Bert then brought everyone up to date regarding the other promotions, "Most important of these, is the promotion of Bob Ventner to the position of Operations Officer. Bob has done a great job, and we are certain his shoulders are broad enough to handle any unexpected events as well. We brought in Hank Hodge to start training for staff positions in year four. Hank is a very bright guy, and he has learned fast. Hank will fill Bob Ventner's former position as Manager of Mechanical Equipment. Back to year four, we also brought in Ray Hilt for training, and he is now an Operations Assistant. Ray will continue in his Assistant position and will be ready to step up whenever another position opens."

"Looking forward to year 27, we will be looking for 4 new college graduates to bring into training status. We will want to be very precise with their training because in the years 40 to 45 we will be passing all of the ship's staff positions to the Generation II personnel."

More problems. In year 33, Hank had complained to Doc Brown about breathing difficulties. Doc checked him over and gave him something to make his breathing easier. Doc then stepped into his office and reported to Betty and Bert, "The problem is his heart. His blood flow is so slow there is not enough oxygen being absorbed in his blood stream. He wants to breathe deeper, but that does not correct the problem. There is a simple operation that might help, but Hank is so weak now, he would not last through the operation. He only has a few days left. Chances are he will die during his sleep."

Betty had a few questions which Doc answered. Then she said to Doc, "Hank is the kind of guy who wants to know what is happening. Do you see anything wrong with me explaining to him what is happening?"

Doc answered, "Nothing wrong with that. I will probably want to know when my time comes too."

Betty said, "I want to be the one to talk to Hank. We will need a few minutes. Please wait here."

Doc and Bert agreed, and settled down to a game of chess, not knowing how long Betty would need. Then to their surprise, Betty opened the door and she and Hank entered Doc's office. Betty took a seat and Hank remained standing. He had something to say, "OK, look, here is what I'd like to do. I probably have a few good days left, so let's use those days. As of right now, I am resigning from the position of Manager of Mechanical Things. Bert, I presume you will slide Ray in as my replacement." He looked at Bert, waiting for his answer.

Bert answered, "Well, yes and …"

Hank cut him off, "Good. Thanks. Then I suggest we have a staff meeting. Right away. Let me talk about me. It's a heart problem, and I could drop dead any time. I have already updated my PDL, so I am not concerned about a turnover to Ray. Are there any questions that the others might have of me? Are there any new bright ideas that we have not had a chance to consider? You know I am good at brainstorming. My brain is just as bright as it ever was. Let's use it. OK?"

Doc spoke up, "Betty, this might be a good time for you and me to leave this conversation." Betty nodded OK. Doc continued, "Bert and Hank, if you want, why don't you stay right here. Betty and I can kill some time in the other room."

Bert called the others – Bob Ventner, Byron Williams, and Ray Hilt. They were there in just a few minutes. Bert opened the meeting, "Gents, Hank is going to run the first part of this meeting," and he nodded toward Hank.

Hank started, "Here is the news. I have had a shortness of breath for some time. Recently it has been getting worse, and now it is bad enough I asked Doc Brown what was wrong. He didn't take long to find out. He said I actually have a heart problem. My blood flow is so low it is not absorbing enough oxygen to carry to the rest of my body. He said my end will probably be during the night. Within a few days." Hank looked toward Bert and said, "Bert, this is probably a good time for your announcement."

Bert took a deep breath, and said, "Ray, I know we are hearing some bad news here, but this is the type of thing we were thinking of when we brought you to the training position. You just heard Hank say he was resigning from his management position. I am now promoting you to that position, and now I need to hear you say you accept the position." Bert leaned toward Ray, as if trying to hear a whisper.

Ray had a very sad look, and said, "Hank, I am so sorry. I was hoping for a promotion, but certainly not at your expense. What can I do to help here?"

Hank took over the meeting. "OK, enough of the sad faces. Ray and I have spent a lot of time in training and turnover. My PDL is up to date. If any of you have questions, this is a good time to ask them." … Long pause … "All right let's just talk. I remember the vibrations problem. We had been on our mission for only three or four weeks. Nick was accelerating the ship, and we were at 5.2 Gs and still accelerating. All of a sudden there was a terrible noise, and we could feel the vibrations throughout the ship. Nick backed off the thrust and as soon as we were back under 5.2 Gs it stopped. Nick sent a message to NSEA for help. We received their response about ten days later. It included a short list of things to check, and if they did not fix the problem, turn around and return to NSEA Campus. We had already checked all of these, but we did it again." … long pause … "Hey guys, this isn't working. Betty and I are going to our bunker. If you have any questions, or anything you want to talk about, call me, or send me a note. I am here to help. But maybe not for a long time."

Hank excused himself and stepped into the next room. He and Betty said their goodbyes to Doc and then returned to their bunker. Hank wasn't feeling too chipper. They had a light dinner and Hank pulled up a good movie they had not seen. Hank passed during the movie.

Hank was 62, and only the second death on the ship. Captain Nick is now 83, and Medic Charlie is 78. Both of these are in good health. We have two retired college professors from Johns Hopkins, both 83, stooped, overweight, and looking like they could drop dead any day. Others on the ship still appear to be in pretty good shape.

Generations IV and V are now in charge and running the community in an effective manner. The Generation VI students are in high school and doing well. The procedures put into place by the Generation

I folks have served the community very well. Our population is now fairly steady at 225 personnel. After four generations we calculated an average death rate of 54 each 20 years. Our old system resulted in lows of about 200 and highs of 250. We then moved away from a target of 54 births every 20 years to an average of 2.7 births each year. This resulted in a larger requirement for teachers, but filling those positions was not difficult. We had plenty of qualified people. And they all had plenty of time to spare.

Our First-Generation staff included 25 workers in the Engineering and Maintenance Department, including personnel with special training for the upkeep of the ship. Such a large staff was no longer necessary, and we reduced the number to five around year 20. The ship's maintenance instructions included schedules out to 25 years. Specific steps were provided for grease and oil to certain items, and five-year replacement for several other items. We were at the 30-year point, and still working with a schedule that ended at 20 years. Our E&M Manager mentioned the problem to the Board and said he wanted to form a committee of mechanical engineers to review the maintenance schedule and prepare a new schedule for the remaining years. This was approved.

The new Maintenance Committee was headed by the E&M Manager and included the maintenance technicians, as well as several degreed mechanical engineers. The mechanical engineers prepared the schedules – grease, oil, and replacement. The maintenance technicians were responsible for carrying out the necessary actions on the schedules. Fortunately, the ship had been provided with the equipment they needed for producing (manufacturing) replacement parts. This was sort of a follow-on to the earlier process called 3-D printing. The process starts with a 3-D scan, which creates a detailed 3-D blueprint. Then the process moves to a modified 3-D printing process combined with a 3-D molding process. The maintenance technicians had some experience using the machinery for simple matters, and the mechanical engineers helped establish some additional processes the technicians had not used before.

The new 3-D process was slow for the first item, while the machinery computer "learns" how to make a new item. Additional repeats for that same item, however, were then quite fast.

The original staff also included Tex Hanson, a bright young man who had been an animal trainer with a circus. He had an exceptional skill for handling recently captured wild animals and turning them into peaceful critters that could be trained. Tex had no college education, but he was very intelligent, and was an excellent helper in many other fields. Somehow, we had neglected to identify someone to follow in his footsteps and be the one to help us know how to handle the wild animals that may exist on

Alpha-1. The Board discussed the matter, and then tasked the Education Committee to locate Tex's PDL and other papers he had prepared for this purpose, and then identify two persons to study Tex's old papers and become our new animal experts.

The Board of Directors scheduled a celebration when we achieved the beginning of year 100. On the first day of the year 100, we gathered in the three meeting rooms (now configured primarily as classrooms) for the celebration. We had talked about a feast, but that idea vanished when somebody said, "Yeah, and what kind of artificial food would that be? Scratch the feast idea, and we will be happy with coffee and cake." The Board quickly agreed to coffee and cake.

The rooms were gayly decorated, with cameras in each room and large screens as well, so we could all see each other. That did a lot to add to the happy mood of the celebration.

At the appropriate hour, our director called for quiet so we could have an appropriate opening from our two church leaders. They had prepared their introductions with a new message. The message reminded everyone of our mission, which was – and is – to search Alpha-1 for new and useful items to take back to Earth.

"The First-Generation folks were committed to this mission, and we need to carry on that commitment. We are now 100 years into the mission, and we need to thank our Maker, for we certainly could not have done this by ourselves. We gather today to give thanks to our God, and we pray again for His continuing guidance in the coming years. Amen."

The prayers left everyone in a serious mood for a minute, but we quickly shifted back to party mood. We had three key speakers who appropriately reminded us of the key events of the past century. Between each of these speakers, we had a short break for refreshments, coffee, tea, and cake.

The next part of the celebration was given to a small group that entertained us all with a short skit, followed by funny stories and jokes. The party was an absolute success.

At the following Board meeting, it was agreed to include the celebration in the Director's PDL. We also agreed it might be better to have the celebrations every 50 years.

A few weeks prior to the scheduled date for the half-century celebration, the Board spent some time discussing other things that might be appropriate. Several Board members thought it would be important to periodically review our mission, and our preparedness to carry out the mission.

One member, however, was opposed to the idea, "Why in the world would we want to start preparing for an event that will not take place for another 250 years?" Another chimed in, "I'm with you on that."

Back to Alice Jefferson, the one who brought up the subject, "Wait a minute. You don't understand. The leaders of the First Generation were picked because they had the right combination of smarts to carry out what had to be done. Further, they had made a commitment to carry out the mission. And they swore to it, and they signed the paper, just below: *So Help Me God*."

"Now calm down, you don't have to get so emotional about this."

"The heck I don't. We, the members of this Board, are from Generations V and VI. Our ancestors from the First Generation committed themselves to carry out a mission. And then they essentially dedicated their offspring – and that includes us – to carrying out their mission for them. That is exactly why we are here."

"OK, I am catching on to what you're saying, and I guess I can go along with it."

"Thank you. Sorry I raised my voice."

The Board President finally spoke up, "I am hearing pros and cons on this. I believe we can take this as a motion to hold periodic reviews of our mission and our preparedness to carry out the mission. We will need a second to that motion, then we can continue our discussion to iron out some details. And then we will take a vote."

The motion was seconded, and then we had a serious discussion. The more we talked, the more we agreed this was something we should do. We decided to have an annual celebration, timed to coincide with high school graduation. The purpose of the celebration would be, of course, to congratulate the new graduates, welcome them to adulthood, and acquaint them with our process for preparing to carry out the mission. Before we were finished, we had to think about who would take care of these things.

Back in the First Generation, the Education Committee was given the tasks related to assigning big brothers and big sisters, and then acquainting new graduates with the PDL process. It is possible the Education Committee might like to take this on as well.

In a joint meeting of the Board of Directors and the Education Committee, it was pointed out that the Education Committee was already overloaded. Steps were then taken to form a new committee, dubbed the Mission Committee. Alice Jefferson quickly volunteered to take on the job as the committee chair. The Board members assisted with a memo to all hands, acquainting them with the new celebration for the graduating high school seniors. The memo also asked for volunteers to join the new committee.

We were averaging 54 new babies in each generation. That works out to be 2.7 each year, and we found that starting a new class for Pre-K every year was not appropriate. We started merging three years, which results in class sizes of about 8 every third year. This worked OK - all the way through the twelfth-grade level,

With that graduation schedule, we then planned a High School Graduation Celebration within a few days of the actual graduation. We had previously held graduation parties, of course, but by including a serious discussion of our Mission, we felt it appropriate to add a little formality to the event. In our first-ever Graduation Celebration, the Master of Ceremonies was our Board President. The overall program started with the President's welcome, and then he introduced the first of our two speakers for the event. The first speaker was the leader of the Education Committee. His talk emphasized the importance of a good education, and how difficult that is to accomplish during our 400-year ride in a spaceship. When finished, the President introduced Alice Jefferson, chair of the Missions Committee.

Alice started by describing NSEA's planning meetings that led to approval of the flight and the very important mission that the team was taking on.

"You, the high school graduates, are members of Generation Eight. Now, just try to imagine our ancestors, 150 years ago, as candidates for this first-ever flight to explore another planet. Some of them had responded to literature that described the upcoming adventure, or they were recruited by the senior staff at the National Space Exploration Administration, NSEA for short. They were invited to a meeting to hear more about this exciting event. The purpose of the mission was to explore the planet named Alpha-1, in search of new resources that may be useful to the scientists and engineers on Planet Earth. The travel time was to be ten to eleven months each way. With two or three months for exploration, the total mission was expected to last for about two years."

"The first meeting was for information. There were many questions, of course, and they were all answered. Some of the answers were: "We don't know" or "Ask us again after we know more." Some of the candidates dropped out. Toward the end of the first meeting, they were informed of the second meeting, which was to be the committal meeting."

"The second meeting was scheduled for one month after the first. All candidates received an email with the schedule included. Candidates were asked to answer this multiple-choice question: Are you committed to join us for our flight to Alpha-1? There were three possible answers: 1. Yes, you can count on me. 2. No, Take my name off the list. 3. Undecided, and I will make my decision at the next meeting."

"Those who answered yes, or undecided, received a letter invitation to the next meeting. Those who answered no received a letter thanking them for their earlier consideration and asking if they wished to participate in similar occasions in the future."

"The second meeting included 215 candidates and started with a question-and-answer period. Afterward, the host asked for a show of hands from those who were withdrawing from the event. Twelve people raised their hands. They were then told to step out and enter the room next door. There they were thanked for their participation and asked to fill out the forms on the table. They entered their names and checked the box for Please take my name off the list. They also indicated yes or no to the question, Do you wish to be considered for similar occasions in the future? They filled out the forms and were dismissed." We didn't know it at the time, but NSEA had planned to cut the list to two hundred. The senior NSEA managers then picked three more to cut from the list and sent to them letters saying they did not make the final selection list.

"The remaining candidates received another briefing on the expected sequence of events for the 2-year mission. The word Mission was becoming more important and was soon joined with the word commitment. Those who were joining for this flight must be dedicated to doing whatever it takes to complete the mission. All those in the audience were then asked to read and sign the forms that were being distributed. The important part was, I, (name), do hereby commit myself to the completion of this mission, whatever it Takes, So Help Me God."

"This had turned into a very moving event for all concerned."

"We launched on Day 1, and our many acceleration drills started on Day 2. All was going well until the 24th day when we experienced some terrible vibrations. We could not find the source; only that the vibrations started whenever we increased our acceleration beyond 5.2 Gs and it stopped when we went below 5.2. We sent a report to NSEA Headquarters, and they directed us to return to NSEA for repairs. On Day 36, when we entered the destination change into the computer, it would not accept the change. Why was that? We discovered that a section of the software had been removed for updating, and the new version was to be installed prior to lift-off. The pre-takeoff checklist showed that the updated section had been installed, but in fact it was not replaced. We, of course, reported this to NSEA, and sent report updates almost daily. But we never received another response from NSEA. Why was that? We may never know.

"So, the question for the day was: what will happen to us if we can't change the destination? Our Ship's Captain and several others researched the program and found the answer: The computer will carry out its instructions to fly to Alpha-1, enter into an orbit around Alpha-1, and stand by for our selection of a place to land on Alpha-1. That part sounded good. The bad part, however, was we cannot accelerate, and our relatively slow speed increases the travel time to 400 years."

"So, what then were the decisions of our First-Generation ancestors? We will do what it takes to complete the mission. *So, Help us God.*"

"That meant they had to pass on the necessary knowledge and instructions to the next generations on how to prepare for carrying out the tasks of the mission. The second generation then had to pass on the knowledge and instructions to the third generation, and so on."

Alice then went on to talk about the role these graduates, as members of the eighth generation, will have in carrying out this mission.

"You will start receiving the necessary knowledge and instructions so that you become experts on these matters. Later, after you actually become the experts, you will take your turn in passing on this necessary knowledge and instructions to those in later generations."

Alice reminded the new graduates they had already been assigned big sisters and big brothers to help them along in these important years of advancing from adolescents to mature adults. This is part of the overall plan.

"As you know, we have encouraged you to follow in the footsteps of your parents, and choose the same, or similar, academic fields. This is part of the plan. That is because it is important that our collective talents are at least as good as those of the first generation."

"You will soon be introduced to the PDL. That is an abbreviation for Pass Down the Line, which is a process for learning what you will need in order to become an expert in your field so you can carry out your mission when the time comes. The PDL process is followed again when it is your turn to pass this on to another generation.

Chapter 5 – Year 399, Time to Slow Down

Back in year 390, we had the computer give us estimates on when we needed to start slowing down. The estimate depended on our level of acceleration – no, make that de-celebration. The method for deceleration is to flip the ship. That entails shutting down the main engines, then using the left-right thrusters to rotate the ship so it is flying backwards. Then we crank up the engines and the main thrusters are slowing down the ship rather than speeding it up. This sounds complicated, but no worry, the computer was pre-programmed for all of these actions, even putting the ship into orbit, and then we can pick our spot for landing.

We picked one G for the deceleration period. Everyone is comfortable with the One-G configuration, and, with the ship flipped, the floors are still floors. The Operations staff keeps everyone informed of what is ahead, so we did not expect to have any surprises, as our ancestors had in Year 1. At one G, the deceleration will take eighteen months. Our landing on Alpha-1 will be in the third month of Year 399.

The Captain and the Operations Officer take turns keeping us informed and give us heads-up notices when unusual things are about to happen. When the main engines are turned off, we are then in zero-G configuration, and if anyone has to move around, they float rather than walk.

The left-right thrusters make noises we had not heard before, and the Ops Officer gives us good warning of that. We are in zero-G for the entire turn-around maneuver.

The Ops Officer was back on the pri-comm and warned us, "We are just about done with the maneuver. The ship is now pointed in the other direction. We will be starting the engines in about a minute, so you might want to hang on to something in case there is any jerking. Wait … wait … NOW." About a minute later, "And … wait … wait … and now the engines are up to speed, and we're almost at one G. The maneuver is over. I'll come back on line if we have any more unusual activities coming up."

Planning. Captain Norman Link had been thinking about this moment for some time. He was quite pleased that the decisions he had been working on in his head proved to be sound. He was wise enough of course to know that he needed to hear the opinions of the officers on his staff, as well as the members of the board of directors. He would be happy, of course, if the final decisions always matched his original thoughts. Most likely, though, some of the decisions will be different; but he would not object to anything that appeared to be the right decision. If others had decisions that were better than his, then he should be happy that he called for the meetings.

He sent a message to his staff, asking them to join in his office for a ten-minute planning session.

"Hi, gang, we need to have a joint planning meeting, consisting of the Ship's Operations Staff and the Board of Directors. That meeting can take place in a few days. Now, this meeting – the one we are having right now - is a preliminary planning meeting. My purpose for this meeting is mostly a heads up, so we can start thinking about the things that are especially important to us. For example, where do we park the ship on Alpha-1? It has to be someplace safe, on a solid surface, and not what might appear to be an empty desert that actually turns out to be a large lake with a thin layer of ice. Anyway, let's get those brains starting to work now. We don't want to be fumbling around next week, and then thinking Gee, I never thought about that."

The Captain's staff members stuck around for a few more minutes with some casual conversation, and then they were gone. The Captain then dialed the number for the Board President, Tom Kennedy.

"Hey Tom. This is Norman. If you have a moment, I need to run something by you."

Tom answered, "Hi Norm. My time is yours. What's up?"

"What would you think about a joint meeting of the Ship's Staff and the Board of Directors? This will be a planning meeting, so we can make sure we have no surprises in the next few months. Our computers will be making most of the decisions for my staff, but I would like to walk through the processes. I have spent the last thirty years preparing for these events, and I believe I know this cold. But I would still like the opportunity for everyone to hear me describe exactly what I am planning on doing. There might be some things that have several options. Which one is the best? Or, when might option A be the preferred choice, and so on?

"We also need to talk about what is important to our experts. We have had discussions about what geologists do. They document the varying types of soil, and rocks. They will create maps of Alpha-1. What else do they do?

"I believe we have plenty of people who are able to start building a city. Should we start making plans – blueprints – for constructing some special buildings?

"We have several people now who are as close to becoming animal trainers as possible. They need to take the lead for checking out the area to identify any animal activity. Does Planet Alpha-1 have any animals we need to protect ourselves against?

"Our scientists and chemists need to check out the water sources – if we have water – to determine if the water is drinkable. They need to check the local grasses, plants, and bushes to see what might be edible. How about allergies? How about poisonous plants?"

Tom spoke up, "Whoa there. This is turning into a long list. And every item you mention will branch off into another ten – or twenty – more items. This is certainly more than the ship's officers and the board of directors can handle."

Norman jumped in, "That's right. I think we will need to meet with everyone, but certainly not all at the same time. How about us starting with your group and mine. We can address our issues, and then we can make plans about the other groups. How about having our first meeting, your group and mine together, say, three days from today?"

"OK," Tom said, "that's great. You can let me know where and when. In the meantime, I'll have a meeting with the board members to get them started thinking in that direction."

Norman replied, "OK, that's three days from now, starting at nine in the morning. I will have the room reserved for all day, just in case. I think that's all we need to do for now. Give me a call if you need. Otherwise, `I'll see you in three days."

Just as soon as he hung up, Norman sent a note to his team to meet this afternoon for an hour or so, for a special planning meeting. They have some pretty sharp people on the Board, and he didn't want them to have a head start.

Both groups had several planning meetings on their own, and then the combined group met as Norm had scheduled.

Norm took the lead and started the discussion with the specific actions that his crew would be responsible for. As he described this, it was clear that he had planned to be responsible for all of the actions. Members of his staff would be his back-up in case of sudden health issues, or, and this was important, in case there were unexpected events that would cause another option to be selected. Jim Anderson, one of the board members, had an interesting question, "Norm, why don't you assign some of the tasks, or events, to your other staff members, and then you could monitor their actions to ensure things are properly done?"

Norm was obviously taken back by the question. "Wow, I think that is the only question I did not anticipate. It is a good question, and please give me a minute while I think of an answer." A minute or so later, Norm addressed the current Operations Officer, Bob Calhoun, and continued, "Bob, have you spent any time thinking about this?"

Bob replied, “As a matter of fact, I have. Now, don’t take this as a final answer, but I am leaning toward what Tom said. With this kind of arrangement, you, as our Skipper. could focus on the big picture, while the rest of us can be the experts for those events assigned to us. These are things we can practice – or maybe rehearse is a better term – at each of these major events. We could identify alternatives, or options, that may be considered. Perhaps we need to write these down someplace, for those *if this, then that*, situations. If any of these are risky, then we may need for our skipper to make the final decision.”

Norm was nodding his head, “I like this. Why don’t we set this one aside for now, and then we can come back to it in a few days. I need to discuss this with my staff. It is apparent they have some great ideas. Jim, my thanks to you for bringing this up.”

Norm took a deep breath and then relaxed, “Tom, I do believe it is your turn.”

“Thanks, Norm. I have several things to bring up. Nothing urgent, but perhaps some changes will be called for since this is a significant change to the atmosphere and lifestyle we have grown accustomed to for the past four hundred years. Before we move to the specifics, I wanted you to know we thought about changing our meeting schedule from monthly to weekly starting about a month prior to arriving at Alpha-1. Then each meeting will have a standing agenda item regarding the schedule.”

Tom placed a few papers in front of him, took a deep breath, and then continued. "The first subject is babies. We don't know what will be facing us after we land on Alpha-1. We will undoubtedly have accidents outside that will need medical attention. Also, we have no way of knowing how to steer clear of poisonous plants. We will probably have encounters with the animals. Some of those might have sharp claws and long teeth. All of these will result in added burdens on our small medical staff."

"Our two doctors and one nurse currently have full-time jobs taking care of our new babies and our soon-to-be mothers. We can't change what is already in progress, but we can ask our young adults to hold off starting to grow their families until we know more about what awaits us outside. Norm, we discussed this to some length yesterday, and we feel it would be appropriate to send a message to everyone - soon, like today or tomorrow - to ask for their full cooperation on this. If we agree, I will take care of the communication. If not, then I won't do anything. Comments?"

Norm commented right away, "That sounds like a great idea to me." Norm turned to the others on his staff, "Do you agree with this?" They all nodded.

"Tom, we agree with your decision on the babies. You earlier mentioned problems we may have with plants and animals. We should schedule separate meetings with our animal experts. Who keeps track of who the experts are, so we can contact them and bring them together to start making plans?"

Tom replied, "We have two people in our Education Committee who know these things. I will have them send us lists of all the experts, according to their specialties. Norm, I believe your staff and the Board members should be participating in all of these meetings."

The next day, Norm called his team together to decide who should do what. The major stages are listed as: 1) Pre-orbit calculations, 2) Transition to orbit, 3) Orbit, taking pictures and looking for a place to land, 4) Landing. They planned a series of meetings, so each of the four staff officers would have a complete understanding of the processes for each of the stages.

Norm and Tom pulled together a team of "animal control" experts for a kick-off meeting. We had five people who had been preparing to take responsibility for these matters. With no information about the fauna of Alpha-1, we came to realize that a group of five people was not enough to explore the surroundings. We decided to join our six geologists, two medics, two computer scientists, a chemist, and a biologist. We called this the animal committee."

"They also formed a team of twenty scientists and engineers to get a head start on building a sawmill and one or two out-buildings, so some meaningful work could be performed outside of the cruiser. This was a very forward-thinking group, and they quickly started efforts to design solar panels, electric motors, drive mechanisms, and blades for the sawmill. Several others were sketching the requirements to manufacture trucks and tractors.

We then had five or six new groups, and they all wanted to meet just about daily for additional preparations. Norm and Tom arranged with Jerry Wyatt, the ship's Manager of the Maintenance and Engineering Division, to rearrange some dividers to create three more meeting rooms. The two basketball courts were now available full-time as meeting rooms in the daytime and basketball courts in the evening.

About two weeks prior to the estimated arrival, Captain Link had a meeting with his team to review the various events coming up. First on the list was to discuss who should do what.

Captain Link described, "The computers will have our ship on autopilot for positioning us 500 miles above the surface of Alpha-1, at a speed of about 17,000 miles per hour. Then we will lower our altitude to 250 miles and start studying the planet and deciding where to land. For comparison, the International Space Station orbits at the same altitudes and speeds. That course will put us into orbit around Alpha-1. If our speed is too slow, our ship can lose altitude, stall, and eventually crash. If we are too fast, we would probably not be able to collect good information on Alpha-1, so we may have to do some trial and error. One reason for this is we do not know the density of the planet's atmosphere. Planet Earth has a core that is largely metallic, with an outer "shell" of rock, soil and water. This density results in a high level of gravity, and we need more speed to stay airborne. As we approach the planet, we will try bouncing radio waves off the surface, and then check our tables to estimate the density of Alpha-1.

Approaching Alpha-1. We slowed to 17,000 mph nearly two days ahead of our expected arrival. Captain Link had designated Bob for flying the approach. Bob obviously could not handle all ten days by himself, so they worked out a four-hour shift arrangement. Bob would have every other shift, and the other three Operations officers took turns at the "off" shifts.

We soon started feeling the effect of the planet's gravity. When our speed increased to 17,500 mph, we adjusted it back down to 16,500. We had expected this would become a simple daily event, but the time intervals between adjustments were growing shorter as we became closer to Alpha-1, so we always appointed someone to have the speed watch.

Two days later, on a Tuesday morning, we were a million miles away from Alpha-1. At this speed we will arrive Thursday afternoon. We are all quite excited. And nervous. None of us had any hands-on training to do any of this. All of our learning so far was from the book – plus some brainstorming with the others. Alpha-1 was just a speck until yesterday. And now we can see dark areas and light areas. Are those areas of land and water, or perhaps clouds? We can only guess.

We earlier calculated Alpha-1 would look about as “large” as the moon viewed from Earth when we were a million miles away. We are closing in at about 400,000 miles per day, and we could almost watch it grow. The images we watched were from cameras installed at various locations, flush with the outer skin of the ship.

How to transition to an orbit was a bit of a mystery to all of us until we spent a few hours thinking and talking about it. In the approach phase we naturally “aim” for the center of the planet. We are careful to keep slowing to our intended speed of 17,000 mph. As our distance to the planet shortens, we start adjusting our “aim” toward the edge of the planet.

We’re in Orbit. We picked a distance of 1,000 miles above the surface to start the transition. At this point, our speed was still about 20,000 mph, with the throttle still at idle. Norm had been mostly quiet during the last several days, but now he could not help but give some directions to Bob. “Bob, I’ll give you the signal in a few minutes to flip the ship. After we are pointed in the right direction, then you can start leveling off. Remember, after the flip, you can point the ship where you want it to go. You should start using the throttle not long after that."

Norm keyed his microphone, "Hello ladies and gentlemen, Norm here. This is what we have been waiting for. We just started our orbit. We are flying level now, and we are about to flip the ship and put the pointed end in front of us. This should not be too complicated, but I want everyone strapped in - just in case. I'll start giving you alerts in a few minutes."

Bob looked at Norm and nodded his head. He was ready.

Norm keyed the speaker button, "Okay, we will execute in one minute. This will be a clockwise spin; a slow spin." A minute later: "Three … two … one … EXECUTE. Our slow spin will take about five or six minutes."

Five minutes later, Norm was back on the air, "That wasn't bad at all. The "flip" is done, and you can unstrap. Our next maneuver will take us down to 500 miles above the surface." Norm turned to Bob and gave him a thumbs-up. Bob smiled and nodded his head several times. He is ready.

Norm gave a few more instructions to Bob, "You can keep the throttle at idle for a few more minutes. Now, tap the nose-down button – very cautiously – There. We are descending at 100 miles per minute. That should take us 5 minutes. And then tap the nose-up to level us off at 500 miles. Now, look at our speed – 18,000 miles per hour. Our planned cruising speed is 17,000, so as soon as we are there, you can increase thc throttle to hold that speed."

A minute or two later, we were level at 500 miles and the speed was steady at 17,000. Norm had been looking at the back of Bob's shirt. It was wringing wet with sweat. Norm placed a hand on Bob's back, "Bob, you have done a great job here. I had thought about breaking your shift into two parts, but you were doing so good I just couldn't change it."

Then to the operations crew, he said, "Bob did a great job. Let's give him a hand." They gave him a good round of applause and a slap on the back, "Yuck, that's wet." Then laughter. Norm said to Bob, "If you had done that any better, I might have to make you the Pilot."

Norm told the crew of the new watch schedule, "Eight hours on and eight hours off. And the two of us will be in the control booth at all times. I figure we will need to do this for about two days, and then we'll look for a place to land.

Norm instructed Bob to go to his bunker, "Get some sleep and be back here in 8 hours to do more."

We are now steady at 17,000 mph, just 500 miles from Alpha-1, and we are feeling the full effect of Gravity. We are very much used to our floors feeling like floors, but now we will feel the full effect of the planet's gravity and we will be standing with our feet on the walls.

Norm decided this was a good time to give everyone another update. Sitting at his desk, he pressed the talk button, "As you can see, we are now directly over Alpha-1, and the view is beautiful. Bob Calhoun has been our primary pilot for the last ten days, and he did a great job taking us to this position."

"We just leveled off at 250 miles above the surface. We are now in straight and level flight. Our purpose is to fly a series of orbits while our cameras take pictures of the terrain, so we can decide exactly where to land the ship. We have a group of experts, including several geologists, studying the pictures, and I will be meeting with them this time tomorrow to hear their recommendations. If all goes well, we should be on the ground two days from today."

"The original captain of this ship, four hundred years ago, was Nick Logan, and he had five successful landings under his belt before this mission was started. Captain Logan left a lot of notes and many recommendations. I have spent many hours studying his notes, as well as the comments of the others who have served as captain during our *Long Journey*."

"After our landing, we will have to sit quiet for several hours while everything cools off. During that period, we will be sampling the air to see if it is safe to breathe. We have a fair-size group prepared for the next phase, which will be to survey the nearby area. Five people in this group have been training in the subject of animal control. They will be accompanied by six others who will be heavily armed – just in case we must fight off any lions or tigers. We hope to find some small to mid-size game that we can use as a peace offering with any larger animals we might come across."

"And we will have a lot more exploring to do as well. Our geologists will be taking notes regarding soil, rocks, and the general layout. Then we will send out another group of medics, chemists, and biologists to determine if we have any poisonous plants. If so, we will need to stay clear of those, then we need to test the water sources, if we have any. We also have a group of architects and engineers who will determine if we can construct a couple of buildings, and a sawmill.

"You might ask, how long will we be here? The original estimate, four hundred years ago, was for a period of two to six months for collecting information before the trip back home. By the way, our computer experts have studied our programs and it appears our trip home may be at a much faster rate; perhaps in less than one year. And now it is time for me to sign off and see how the geologists are doing. I will talk to you again in the morning. Good-by for now."

In the Captain's office we had our four operations officers, as well as Tom from the Board of Directors, and three geologists. Melvin Sherwood had taken the lead for the geologists and was carefully telling us how to read the maps, how to distinguish between land and water, and so on. At 17,000 mph, we took an hour and fifteen minutes to make a complete orbit. Simple math, then, tells us the circumference is about 21,250 miles. Compare that to the 24,000-mile circumference of Earth, and we see this is a somewhat smaller planet.

Our pictures gave us a good view of the area within five hundred miles to the left and right of our flight path from about two-dozen orbits. Alpha-1 receives light from two sources. Melvin said we could probably trust what we see under either of the light sources but should disregard the pictures taken during darkness.

We started ruling out the areas we definitely could not use as a landing pad. Water areas were out for obvious reasons. That rejection took care of nearly half of the total area we were viewing. Mountainous areas were out, as well as other rough terrains. Forest areas were out. We quickly rejected about ninety percent of the total area. We then increased the picture size for the remaining areas by four times and were delighted with the detail. We picked out three areas that looked like reasonable choices. We decided to make a low pass – at an altitude of fifty miles – in the morning for our final decision. We would also look at two areas where we saw several small lakes. They might be useful as water sources.

Next Morning, Low Pass for a Better Look. In the morning, Captain Norman was in the driver's seat. He hit the talk button and said, "Good morning, everyone. I now have on my Captain's hat, and we are preparing to land the ship. But we have one minor change. Last night we identified five possible areas for landing, so we will make a low pass for each to help make our final decision, and then we will land on the next orbit. The cameras are set to give you a view in all directions, so make sure you are watching. 'Bye for now."

Before quitting last night, they entered the five locations into the computer, so Norm could focus on altitude and speed. The computer and the cameras would do the rest – for this pass. The Captain reduced power by five percent and nudged the nose slightly below the horizon. When the altitude was one hundred miles, he adjusted power to steady-up at 12,000 mph.

In the morning, he descended fifty miles and started reducing speed. The first possible landing area was scattered with large boulders, so that wouldn't work. The next area had several lakes – remember this if we need water – but maybe too hilly. The third area looked good for a landing spot, and also had several lakes. The fourth and fifth were completely covered with trees. No place to land. He added power to go back up to 250 miles, so he would have more maneuvering room.

Norman turned to the others in the room, “Hey guys, I like the third area for landing. Do you agree?”

Melvin checked with the others, and they all nodded. Then Melvin replied to Norm, “Good news, Norm, we all agree with you.”

Final Pass and Landing. Again, Norm’s work on the computer last night was very worthwhile. He inserted into the computer an action list for starting and stopping the various elements of the maneuver. The computer flashed a message, Confirm Area 3. Norm pushed the Yes button. The next computer message was, Throttle at Idle in 1 Minute, and then the computer showed the countdown, 1, 2 . . . 58, 59, IDLE NOW. Norm moved the throttle to idle and started pointing the nose upward. The computer flashed, Throttle up NOW. Norm immediately moved the throttle to the 50% level, and then continued slowly moving it toward 100%. Norm was watching the speed indicator. When it indicated zero, he held the throttle where it was: 83%. At this point the ship was about 100 miles above the surface, and motionless, with the engine’s thrust in one direction exactly equal and opposite to the gravity in the other direction. Now showing on the screen in front of Norm was a live picture of the landing area, with a circle superimposed at the exact center of the area on the ground they had selected. Also on the screen was a small "x" which indicated the spot where the ship is currently headed for the landing, taking into consideration the altitude, speed, and current ship attitude. There is a toggle switch in front of Norm for making north-south and east-west adjustments to the ship's path. Norm was pretty much on his own now. The x was left of the circle, so he tried two taps to the right. The ship responded and Norm watched x moving to the right. A minute later the x stopped in the center of the O. Norm was happy with that, but he was sweating, and his hands were wet. Bob saw this and handed him a small towel.

The next step was to start down, which Norm initiated by reducing power from 83% to 81%. The speed indicator moved from zero to 5 mph. The altitude indicator showed we were 92 miles above the surface. There was also a time gauge to show the remaining time to landing. That was simple math by dividing the distance by the speed, 18.4 hours, but Norm said it was very helpful to have that in front of him. A few minutes later we still had 88 miles to go. Norm pulled the throttle back to 80%. A minute later the speed was 16 mph, and the remaining time was 5.5 hours. After several more of these very cautious adjustments, our altitude was 61 miles, the speed was 23 mph, and the time to go was 2.6 hours.

Norm saw that the "x" had drifted about a mile north of the circle. He tapped the toggle twice, and then leaned back for a moment. "Bob, can you find another towel for me?" Bob replied right away, "Here's one." He handed Norm a towel and showed him his stack of a dozen more towels.

At 35 miles, Norm said to the group, "We have an hour and a half to go, so I am starting to slow down." He bumped the throttle from 70% to 75%, waited a minute and went further to 80%. He dried off his face and hands, and then waited for the ship and the gages to settle down. The altitude was now 30 miles, our speed was 18 mph and the time remaining was 1.7 hours.

Wondering how much time he had before he would lose the light, Norm checked his wristwatch. Somewhat embarrassed (to himself only), he asked Melvin how much time we had before dark.

Melvin grinned and replied, "I saw you looking at your watch, so I knew you would be asking. Sunset will be about two hours after we land."

Norm commented, "I was hoping we could step outside for a breath of fresh air. But that's not enough time for things to cool off. Besides, our exhaust will probably spoil the freshness for at least that long.

Now, with one hour to go, we were at 24 miles.

At 20 miles, Norm added some thrust to slow the ship down to 12 mph.

At 10 miles, he moved the throttle to 81%, and the ship slowed to 5 mph.

At one mile, Norm slowly added throttle until the ship stopped. In midair. That was sort of to prove to himself that we really could stop. That done, he squeezed the throttle back to 80.5%. The altitude changed slowly. A half-mile, the x was in the center of the circle. Right on! One tenth of a mile. Half a minute later we heard a thump, and we were there. Our four-hundred-year *Long Journey* has come to an end. Norm secured the engines.

Those in the Operations Office went crazy! shaking hands and hugging each other. Some had tears on their cheeks. It was a moment they would remember forever.

We have Landed. Norm sat back down and pushed his talk button. “Ladies and gentlemen, WE HAVE LANDED! Now, I hate to be a prude, but nobody should step out of the ship until tomorrow morning. It will be dark in another hour, and that’s not enough for the engines to cool off, or for the exhaust fumes to clear out. In the morning, about an hour after sunrise our first crew will go outside and check out the area. After I have their reports, I will be back on the air to talk about other events for tomorrow.”

Some other work was ongoing. The geologists had set up a temporary meeting room on one of the lower levels. Jerry Wyatt arranged for the ship's engineers to pipe in samples of the outside air so it could be tested. From our distant ancestors to us, we all have grown accustomed to breathing very safe air on earth, which is 78% nitrogen, 21% oxygen, and trace amounts of several other gases, all of which are safe. The scientists were happy with the new samples, and Clyde Holdren, our senior scientist, made the call to Norm, "Hi Norm, we have good news. The sample air showed 74% nitrogen and 24% oxygen. That is a little better than we had on Earth."

Norm asked Clyde, "That's great. Is it OK then to start circulating air throughout the ship? We are all anxious to breath some of that fresh air."

Clyde replied quickly, "No, that was just the easy part. We still need to test the trace gases to see if any of them are harmful. There are five or six that we need to check for, and we should have the final answer for you within the hour."

"OK, thanks. I will look forward to your call."

An hour later Norm answered the call from Clyde, "Norm, I am pleased to inform you that we had seven trace gases, and all seven have proved to be safe. I am even more pleased to report that three of us have been breathing the air for the last half-hour and we are still alive and kicking."

We are all in a good mood this evening. Norm had started reading a book but could not keep his mind focused on the story. He was thinking this would be a good time to pop a bottle of champagne. “I think I’ll go out looking for grapes tomorrow.”

Awake early in the morning, Norm started mentally running down his action list, and looking for things he might have overlooked. “I don’t have time to look for grapes.” Counting from the ground, the first thirty feet of SC12 is taken up by the engine room. That takes up much more space than other cruisers because several parts are new design or new technology. Some of the NSEA senior executives were not convinced that everything would hold up for the two-year mission. Huh! How about a four-hundred-year mission? So, we have a complete set of new technology equipment, and the older equipment as backup. Oh well.

We have some heavy equipment to take outside. The current staircase is too narrow, so we will take those out through the double doors on the first floor, and carefully lower them with rope.

First Trip Outside. Norm joined Melvin, his senior geologist, and Billy Stone, head of the animal control group, and the rest of the team, were on the first floor and preparing to go down the narrow staircase for their first outing. They were all wearing mikes and earphones. Everyone had mikes on so they wouldn’t have to search for switches, in case of need.

This, the first group, included five designated animal control "experts" and six bodyguards, in case we ran into any of those large animals with long teeth and sharp claws that Tex was always referring to. They were all serious, with only an occasional bit of laughing or carrying on. The group moved down the tight stairs and then outside. The first comment was, "Hey, this looks a lot like home," and another, "Some of the plants have funny-looking leaves. I've never seen any like these in those catalogs I studied."

While taking positions before starting their walk, one member of the crew reported seeing a small animal, "That looked like a fox. Over there," and he pointed in that direction so the camera people could point a camera to that spot, then they all lost contact with the fox. They were being cautious, and took their positions, with five people stretched out in a line, with them spaced about twenty feet apart. Billy Stone, the senior animal expert, was in the outermost position, about a hundred feet from the ship. After they were all in position, they started to swing their "line" around the ship, like a spoke in a bicycle wheel. After a few minutes, one of the cameramen said, "There it is, and yes, it looks like a fox, just beyond the last guy out there. A minute later, he said, "Look out, I think he's preparing to charge."

Two of the bodyguards were taking positions around Billy, and then they saw the fox closing the gap on Billy. Both bodyguards fired, one with a rifle, the other with a shotgun. The fox was dead. Very dead. Everyone had to gather around to check out the dead fox. As near as they could tell it looked like an ordinary red fox. Henry disagreed, "I believe this guy is larger than the foxes back on Earth. Perhaps this is a wolf, or some sort of mix of fox and wolf. In any case, I would want to stay away from those teeth."

One of the bodyguards grabbed the dead fox/wolf by the tail and dragged it to the ship, a couple of yards from the door.

Back in position, the animal guys finished the circle. Then they repositioned out to about double the distance as before. About twenty minutes later, Billy, still in the outer-most position, stopped, waved his hands over his head, and pointed out another animal. Billy, now looking like he was frozen to the ground, put a finger to his lips, to tell everyone to be quiet. Then he turned toward whatever he was looking at. The cameraman spoke, "Be careful, that looks like a bear."

Billy was still stopped, and two bodyguards joined him. The bear was quiet, but his eyes were focused on these strange, alien creatures. Billy spoke up, quietly, "I think we should leave the bear alone. But post one of the bodyguards here in case he goes crazy on us. I will start the rest of the team moving around the circle again."

The cameraman told the group, "I'll keep a camera on Mister Bear, and I'll let you know if there is any action there."

It took another hour to rotate through the large circle. The team had been outside for two hours now. They had some tense moments, and everyone looked worn out. Norm and Tom came down to the "first floor" to join for the debrief.

The first action was to post a sign, saying, CAUTION, THERE IS A BEAR ABOUT 100 YARDS OUT FROM HERE. STAY AWAY.

Billy said to the group, "Who knows how to skin a fox? I would like to take a present to Mister Bear, to see if we can be friends. And I would like the bodyguards to provide cover. And I'd like to have a handgun in my pocket."

Two hours later, Billy started a slow walk toward the bear. In his left hand was a juicy-looking piece of fox. In his right was the handgun. Remembering Tex's notes from four hundred years before, Billy was doing his best to act normal, and not scared. And he tried to sound normal, "Hello Mister Bear. I have a nice fresh piece of meat for you. I am sure you will like this. Can we be friends?"

Billy had thought he might walk right up to Mister Bear and hand him the meat, but his courage level was not running that high. He finally held the meat up to his shoulder level, jiggled it a bit, and then tossed it the remaining twenty feet. Mister Bear let out a snort. Not a growl. Perhaps he was sniffing the fox odor. Mister Bear slowly uncurled and stood up. Billy then saw the two cubs that Mister – no, Mrs. Bear - was watching over, and he spoke quietly to his hot mike, “There are two bear cubs. Mama Bear looks OK, but this could get dicey if Papa Bear comes back soon. I’m turning back now.”

Billy wanted to run, but Tex had warned, if you run you will be chased. Billy walked slowly back to the ship. The bodyguards joined him a couple of minutes later. They came inside, closed the door, and ran up the stairs. The whole group was still there, and now joined by about two dozen others who wanted to be closer to the action. They applauded Billy when he stepped into the room.

Billy grinned, took a bow, and then started jumping up and down. “I have to settle my nerves,” and then he leaned against the wall. Billy took a deep breath, started looking relaxed, and then continued, “We need to carve up the rest of the fox meat. It looks like we can expect another three or four usable pieces. Then we need to form a hunting group to gather more meat for the Bear Family. Tomorrow morning, I would like to give the Bear Family another treat. A little larger piece this time, please.” A minute later, Billy had another thought, “But why limit this to me? If one is OK, then two should be better. Who would like to go with me tomorrow?”

George Thompson, another member of Animal Control, said he would like to do this. No one argued with that, so Billy continued, “Let’s gather at eight in the morning.” Heads nodded to indicate agreement. Billy went on, “I think we should keep these events within the animal control group. When George and I are scheduled for a feeding event, I would like our bodyguards to be from our group. Also, this afternoon, I would like you guys, and Sally Blanchard, to go on a hunting trip. Foxes are fine. Wolves and raccoons should be OK. Deer are OK, Squirrels and rabbits are OK. No skunks and no snakes. And no lions or tigers. And no bears, unless you are attacked.”

Outside Staircase. There is a large door that opens to the outside on the first level above the engine room. That is thirty feet above the ground. Or maybe a foot or two less, as the ship may have settled a bit. Our only usable exit right now is a small staircase that spirals down the outer wall of the engine room which opens with a four-foot-wide door. We need an outside staircase so we can exit and enter at the level above the engine room, where we have double doors, with an eight-foot total width.

Norm and Jerry had talked a few months earlier about building a wooden staircase on the outside. Jerry prepared a sketch of the staircase they planned to build. The sketch had a lot of detail, and Norm saw they were planning to use fresh-cut logs in the construction. That would be OK for the basic frame, but we will need boards, not logs, for those steps.

Sawmill. We need to expedite building a sawmill so we can cut the boards we need for the steps. Jerry's engineers also were planning for a small group of buildings, one or two of which would be for the sawmill. They manufactured two heavy-duty drive mechanisms for the large saws. Solar panels will be placed on the roofs of the buildings, and we will have a converter to change the circuit from DC to AC. Some of these items had never even been thought about in the past, but as Jerry would say, "We can learn anything we need from the Internet. Jerry and his two teams were going to be very busy.

Plants and Water. Norm and Tom were anxious for the small team of doctors, chemists, and biologists to start checking the area for what was OK and what was not. We are certainly aware that the natural flora on Earth has a few poisonous plants and an array of other plants that can make life miserable for those with allergies. We would like to have a source for drinking water, but it will need to be carefully tested to ensure it is drinkable.

We scheduled a brief outing for the team later today and warned them to stay away from the marked area where the Bear Family lived. Also, they were not to wander more than a hundred yards from the ship. For safety purposes, they were accompanied by three armed bodyguards.

Tom added this, "One more thing. Please be back inside the ship before dark. And let's have a debrief on the first floor. Norm and I will be here waiting for you."

Two hours later, they gathered on the first floor. The meeting was brief. There was no water source to check out. That would have to wait until they could venture out to a river or lake. A call to Melvin told them there was a small lake about five miles away. What direction? According to our magnetic compasses the direction was five degrees north of northeast, or 040 degrees. Where is North? Back on Earth, there was a magnetic "attraction" very close to the North Pole, so the needles on our compasses were labeled with an "N." There is definitely a strong magnetic attraction here on Alpha-1, but we don't know where that is. We decided that didn't really matter for now. All of our compasses point in the same direction, and we can live with that. It will be several more days before we are ready to venture out that far.

Plant samples had been obtained and placed in plastic bags. The scientists would need several hours in their laboratory to do the necessary testing. They promised to send reports to Norm and Tom, and then they would post the samples on the first floor, so all could review before exiting the ship.

Several hours later, Clyde set up a conference call with all those who were involved with today's activities, including Norm and Tom. After some introductory comments, Clyde started his report, “I have some good news, and some other news that is almost good. Our first task was to check for poisonous plants. This took a while because we had thirty-two plants to test, and a dozen possible poisons to look for. The good news is that we have no poisonous plants. That means you can eat anything you want, as long as you boil it first. That will take care of any itchy stuff that you don’t want to put in your stomach. After all that, I still recommend we not eat any of these. Besides, none of them look very appetizing.

“The other category includes plants that can cause a rash or itching. We found one plant, a vine, that appears to cause a reaction like poison ivy. But - and now please pay attention to this - the leaves are different, so make sure you look at the samples we will be hanging on the wall, probably right after we finish this call.”

There were a few questions, which Clyde answered easily. Tom thanked Clyde for his work. Norm said, “I believe this is the perfect ending to a very successful day. We will have to remember this as DAY ONE of our visit to Alpha-1. Now let’s all get some important rest and have some happy dreams. Then they all joined in for a rather loud “GOOD NIGHT.”

Animal Control. The next day, early in the afternoon, four members went on a hunting trip. This was intended to be a daily event, so they figured out a way to hunt in new areas, as best they could. For today, they would hunt in the northern quadrant, starting at a hundred yards out from the ship. The next day they would be in the eastern quadrant, then south, and then west. After four of these, they would move out to two hundred yards, and then so on for later days.

The afternoon trip was successful, and they brought back the remains of two foxes and a deer. Our cooks removed the meat and started saving these for a cookout, perhaps the next day.

Sawmill. After they had provided the staircase workers the boards they needed, the sawmill team started on boards for their temporary buildings. They had picked two designs, and the most important considerations were expedient, useful, easy to build, and easy to change. The small buildings were six feet by six feet, with four posts and a roof. The other design was nine feet by twelve feet, with nine posts, and a roof. Walls and floors would come later. Knowing their tasks would be ongoing for several more days, Jerry dismissed his team an hour before dark, so they could rest up and prepare for the coming days. Many of the members of this team stopped to view the other team's work on the stairs. Several of the younger men decided to help out.

The main part of the sawmill consisted of two of the larger designs, side by side, with the common wall having a twelve-foot dimension. This building contained two large saw tables. All four nine-foot sections had double-wide doors so milling long timbers was not restricted to the length of the building. Attached to this construction was a small six by six room, primarily for the converter and batteries, but also for spare parts and additional equipment, such as two battery-powered circular saws, and the manual long blades that are necessary for taking down very large trees. We had a dozen workers, including ten men and two women, working on the buildings, so this work was going fast. The work was interesting, so we usually had another dozen or so spectators watching the action.

We started out believing the senior carpenters should be mature men (or women), and assistants would be picked from those in the 16 to 20 age group. We soon realized, however, that no one really had any experience cutting boards or pounding nails, so we quickly changed our ways to ask for some number of volunteers, and then we let them work out any seniority arrangements on their own.

On the other side of the ship, we started a small "community" of buildings. Most of these were in groups of four buildings, each one with the basic designs, again starting with posts and roofs. As soon as roofs were in place, technicians moved more of the solar panels from the grass to the roofs.

Some of the younger folks were asking for an outdoor basketball court. Surprisingly, others said, "We already have a nice basketball court inside. How about a football field? Or baseball?"

Tom asked Norm what he thought about that. Norm offered, "Tell them that might come later. In the meantime, they should grab a hammer or shovel and join the construction teams."

A few days later, in a conversation between Tom and Norm, they decided the construction should probably end for now, before everyone started wanting to move from the ship to the buildings. We need a longer-term plan to decide just what we are going to be doing, and how long we are going to be here. Tom suggested, "I believe you are right on that, but we still need to find something to keep the younger ones busy. Maybe an outdoor basketball court is not a bad idea after all."

Board Meeting. Tom and Norm worked together to send a message to schedule a board meeting in a few days. The agenda would include looking for a water source, scoping out our area to at least ten miles, maybe more, revisiting our mission, and preparing a temporary schedule.

Outside Staircase. Engineers for this project were up at the break of day, anxious to have this ready for everyone to use so they would no longer have to squeeze up the stairs going through the engine room. We were not allowed to fasten anything to the outer skin of the ship, no nails, no screws. Their design was to position the top part of the frame to be level with the ship's first floor. They drilled holes in the sturdy frame pieces at the sixteen-foot level, and then connected angular braces to the left and right sides. These helped to make the whole frame rigid. At the bottom, they were fastened to wooden anchors, buried in the ground. Near the top of the staircase, they fastened a heavy rope at the top left corner of the staircase frame, stretched the rope tightly around the ship, and then fastened it at the opposite side of the frame.

In the early afternoon, with the frame finally in place, and fully tested for sturdiness, they could then focus on the steps. By this time the sawmill had been creating 2x4 and 2x6 boards out of fresh-cut trees, and the staircase team now had a good supply of new lumber.

They had initially planned to quit working before dark, but now, with the end in sight, they unanimously agreed to keep going. Jerry brought out several bright lights and extension cords to provide the light they needed. Finally, two and a half hours after dark, they were far enough along to call it quits for the day.

The next morning, the steps were double-checked, primarily for safety purposes, and a few minor errors were taken care of. Railings were next and were to be completed in the mid-afternoon. Tom and Norm were then invited to be the first to climb the "finished" staircase. They congratulated the team on their fine work, and proclaimed, "This project is done."

Animal Control. Billy, George, and two bodyguards met at eight in the morning, on the first floor. Billy had visited the kitchen several hours earlier to pick up his allotted one-half fox, so the Bear Family would find the meat at a normal temperature, and not just out of the freezer. When they reached the ground, Billy was finding his nerves were starting to act up. Every few minutes, he would stop and jump up and down a couple of times, The first time this happened, they laughed at him. But after his angry look, they decided to ignore it. After all, maybe he was doing the best he could.

A few minutes later, the Bear Family was in sight. Billy had been carrying the fox meat in a bag. Billy gave a "stop" signal to his small group, and then handed George the fox meat, from the bag, "Here, it is your turn."

George saw that the bears were looking at him - or were they looking at the fox meat in his hand?

George looked flabbergasted, “What? Why? I don’t know what to do.”

Billy spoke very calmly; calmer than before, “It’s easy. Hold the meat in front of you so the bears can see it. Take three or four steps, then toss the meat to the bears. If you are OK with this, we can take turns feeding the bears.” George started feeling his nerves reacting, and thought to himself, “This is why Billy wanted me here! “

Papa Bear picked up the fox meat and carried it back to their lair. Mama bear pulled out two pieces of meat to put in front of the kids. Papa Bear ate the rest. No wonder he was nearly twice the size of Mama Bear.

They turned, walked a few steps, then stopped. George looked like he enjoyed that, and he said to the group, “That was exciting. How about the four of us taking turns feeding the bears?” Billy looked disappointed. The two bodyguards looked shocked.

One of the bodyguards said, “Count me out,” and the other said, “Me too.”

Billy looked worried, but before he could speak, George smiled and said, Well Billy, that leaves you and me. Your turn tomorrow.”

The three of them looked at Billy, who then said, very quietly, “Yeah, right.”

As it turned out, Billy left the team and George took over the daily task of feeding the Bear Family. George dismissed one of the bodyguards, and eventually the remaining bodyguard no longer went the whole distance. George felt OK as long as the bodyguard was within shooting range.

Billy then took over as head of the larger team for checking out the surrounding areas. For the first several days it was fairly easy for a five-person team to check out a quadrant, covering the area from two miles to three miles. But the next section required a team of ten, and the one after that, covering from four to five miles, would require twenty.

Tom and Norm were studying the draft map that Melvin was keeping up to date. Norm recalled there were pictures showing several lakes a few miles away. Melvin enlarged these pictures another four times so they could have a closer look. The closest lake was almost due west from here, actually 275 degrees on the compass, and there were no steep hills or mountains. Almost all of the area is covered by trees, so they were not able to see details at ground level. Someone pointed out we might come across a stream on the way.

We figured it would take at least one whole day to walk twenty miles. Walking through undeveloped woodsy areas would be quite slow, probably no more than two miles an hour. They talked about equipment and supplies we would have to take. Tents: maybe.

Sleeping bags: definitely. Food: sandwiches, artificial veggies, and rifles for hunting fresh meat.

They decided to call a special Board Meeting so they could find out any helpful ideas from the others. As expected, the interest was very high, and the board was very much in favor of a special trip to the lake. Billy was eager to work off his previous reputation, but the Board was not about to trust him as the leader for this important assignment. Patricia Arnold volunteered for the lead. She was in training as a member of the cruiser's operations staff and had impressed others with her leadership skills and her nerves of steel. She had also spent three days searching the new areas.

Chapter 6 – Searching for Water

A team was formed, with Patricia as the lead, and others included Tom, Norm, Melvin, and Billy. During our meeting, we added three others as bodyguards: Henry Jackson, David Blake, and Richard White. Everyone would carry their own sleeping gear, sandwiches, and eating utensils.

Day 1, Walk to First Lake. The team members joined at six in the morning, at the bottom of the stairs. Ten minutes later they decided they could see well enough, and off they went.

The first hour was quite easy. They were all young enough and healthy enough to enjoy the outing. They were definitely having a good time, with lively hops and skipping in the open areas, and then definitely more careful around areas with down branches and heavy overgrowth. After four hours, however, not even halfway for today, they had slowed down quite a bit. Pat said this was a good time for a break. Everyone found a place to sit. One laid down on the ground, but quickly jumped up when someone said, "That ground might be covered with ants and spiders."

Patricia Arnold said to the group, "Come to think of it, I have not seen any ants. Has anyone seen any ants? Or spiders? This is information we should be keeping track of. Let's start keeping notes, and after our trip I will combine your notes and give a report to the whole crew."

Then she placed a call to Clyde Holdren for a sitrep, "Hi Clyde. I think we are a little older than we thought." Pause. . . "Oh, you're right. We are not older. We're just out of shape. Here's an interesting item. Lucky us, we have not seen any ants or spiders on the trip. I am wondering if we have them at all, on Alpha-1. We'll keep our eyes focused for ants and spiders, and I'll let you know more about that later. Oh, one more thing. You said sunset will be around eight. I think we will quit for the day around six. We should be at the sixteen-mile point, and we will need a few minutes to do some hunting. We should be at the lake around noon tomorrow. I will send you a text report later today, and then on arrival tomorrow. I believe text messages will travel OK, but voice may not. If I find different, I'll let you know."

Tom and Norm looked at each other, smiling. Tom said, "I think we picked the right person as lead for this trip."

Norm replied, "I agree. And that's exactly why we picked her for the lieutenant position last year."

Patricia called for everyone's attention, "Here's what we're going to do. We'll be back on our feet in a minute. Let's try to keep going until six. Then we will stop for the evening. We all brought sandwiches for both lunch and supper. The sandwiches are not very appetizing, but they are healthy. We will have time for some hunting in case anyone is interested. Any questions?" Silence. "OK, backpacks up, and away we go. We will have ten-minute breaks every hour.

The first hour passed quickly, but the break was welcome. One member of the team, Henry, had a question for Pat, "I have seen several small animals. We are all carrying rifles, and I thought it might be nice if we could shoot something for supper. So, how do you feel about some of us being designated as hunters?"

Pat replied, "Good idea. We will have a break in a few minutes. We can discuss it then."

Fifteen minutes later, Pat called for their afternoon break. The backpacks came off quickly, and more than half of the team disappeared for a few minutes, for nature calls.

When everyone returned, Pat brought up the hunting subject, "Henry asked me a question a few minutes ago about hunting while we are walking toward our destination. I think that will be OK, but we need a few rules. I like being the leader for the team, and I like to be at the front of our group. I would also like for the hunters to be within my sight. So, let's say we can have two in a hunting mode. One out to my left and one to the right. Stay within fifty yards of me and remain in sight at all times. You are cleared to fire in the ninety-degree area from dead ahead to straight out beyond the side you are on. Oh, I forgot to ask; How many of you would like to try hunting while we are in walking mode?"

Two hands went up right away. Then there were two more, and then they were all up. All except for Pat. She grinned, and said, "Well it's not that I wouldn't like to do a little hunting. But I can't be a hunter and the leader at the same time."

Tom spoke up, "Pat, I can switch places with you for a while. Will that work?"

Patricia answered quickly, "Yes, that can work. Thank you."

"OK, we will stop for a break every hour. We can start out with Henry and David as designated hunters on the first shift, Norm and Richard on the second, Tom and Melvin next, and then Billy and me on the fourth shift. Is this OK for everyone?"

There were a few questions. "What are we hunting for?" Laughing. Several joined in for the answer, "Anything edible." More laughing. "No skunks."

"One more thought; If we run out of time today, then we continue tomorrow morning. Hunters in the same sequence. And now, it's time to put on the backpacks and start moving along."

Pat commented, “I suppose our group looks like a moving ‘T’ from the air, with Pat in the lead, Henry and David in the hunter positions, and the other five in single file.” The two hunters were carrying their rifles in front, sometimes using one hand, but usually with both hands. The rest of us had rifles too, but they were strapped to our backpacks. We were armed with 30 caliber rifles. Just right for deer or fox, but a little heavy for squirrel.

Perhaps fifteen minutes had gone by, and we saw Henry stop. He had his rifle in the “ready” position, moving slightly, trying to get a clean shot at that squirrel in the tree just ahead of him. Finally, he pulled the trigger, BANG, BANG, twice. The squirrel fell to the ground. Henry picked it up by the tail. It was dead, so he stuffed it in one of his plastic “game” bags, and attached it to his backpack, so it hung down, as a trophy, and Henry’s first conquest.

We were stopped only a couple of minutes, then Pat used her loud leader’s voice to holler, “Backpacks up!” and then, a few seconds later, “Let’s go!”

A half-hour later, we stopped for another short break. Pat positioned Norm to the left hunter position and Richard to the right.

After another half-hour, Richard motioned to Pat for everyone to stop, be quiet, and stay low in the weeds – out of sight. Then Richard bent way down and appeared to be studying something up ahead. He eventually pointed toward his target, a yearling buck. One year old, with its first set of antlers. The deer was almost a hundred yards away, and Richard wanted to sneak up to maybe half of that distance. The group participated by not talking and staying low, and, hopefully, out of sight of the deer.

Richard disappeared. Then we saw some of the weed-tops waving, and we knew Richard was closing in on the deer. We remembered our cookout, about a week earlier, and our first-ever taste of deer meat. The cooks called it venison. We didn't know why, but that's what it's called. Several of us looked it up on the Internet, and yes, that is the correct term.

Then the weeds were still. All was quiet. We saw Richard's head reappear. The rifle came into view. Richard was waiting for the right moment. Then we heard, BANG, BANG. Two shots. Richard stood up and yelled, "Yahoo! He's down. Wait for a couple of minutes while I make sure he is down for good."

Two or three minutes later, we saw Richard raise the rifle for one more shot. Then he motioned for us to come forward. He said, "The deer looked dead, but I wanted to make sure. The last thing we want is for him to wake up and try to fight his way out of this."

We picked Henry, Richard, and Tom to carve out the edible parts and place those in three plastic bags. The three carvers would also be the carriers. The designated cooks for a pleasant meal this evening were Melvin, Billy, and Pat. Norm and David would dig a hole and bury the unusable animal parts.

Eventually we were ready to move forward. The time was almost three-thirty, and Pat said, "If we leave now, we still have two more hours for our hiking. Tom, you are the hunter on the left, and Melvin is on the right. You can pick up your backpacks and take your positions." Then, just a few minutes later, she grinned and said, "Forward March," in her best gruff Marine voice, as she remembered from a good movie placed in the Vietnam War era.

The next two hours passed quickly. At a few minutes before six, they picked an area under some large trees, hopefully good protection in case of rain, and close to an open area that should be safe enough for our campfire. While everyone else was busy, looking for a good place to unroll their sleeping bags, the cooks were making their preparations for dinner.

Venison was to be the main course, but what else? The packed lunch bags consisted of bread, artificial veggies, and small packages of ketchup and mustard.

They started a fire, using a good amount of wood for flame, and about twenty rocks positioned at the perimeter of the fire. They let the fire burn down to coals, pulled the hot rocks closer to the center, and positioned the pieces of venison over the fire, held up by an arrangement of green branches for spits. Dinner was served to each on plastic plates with a nice piece of venison, two slices of bread, and a side-dish salad of artificial veggies.

Everyone enjoyed their dinner. We stayed in a circle around the fire and talked for an hour. Subjects ranged from jokes and funny stories, to what we might find at the lake tomorrow, and to completing our mission. Then we talked about returning home, to Earth, and will we ever have a chance to join our very distant families – our ancestors. That part of the conversation became serious, and several of us were in tears.

A few minutes later, the mood changed, and we sang a couple of silly songs. Then we were quiet again and Pat said we still have three or four hours of hiking tomorrow, and it was time for us to turn in. The fire had burned down to red coals. We crawled into our sleeping bags and said good night.

An hour or so later, we heard, “WhOOoo.” Several of our group spoke up at the same time, “What was that?” Then someone else said, “That’s just an owl. He’s looking for bugs, or mice, for his dinner.” We went back to sleep. Someone said later he dreamed about giant birds carrying us back to their nests, to feed their babies.

Day 2. At the first break of twilight, several of us heard something moving through the grass. Was that a snake … or a mouse … or what?

Henry, the brave explorer, crawled out of his bag, and stood up. We heard a scurry of tiny feet, and Henry turned on his flashlight. Then he saw them. "I see two foxes. Baby foxes. Cute as can be. But I believe we should leave them alone. Mama Fox may be nearby, and she'll raise a ruckus."

Then we heard a strange sound. It was sort of a bark and a growl, mixed together, but not an angry sound. Henry moved his flashlight around and saw a full-grown fox, patiently waiting for the naughty kids to follow her back to their den. Then they were gone.

None of us were now in the mood to go back to sleep, so we took this as our reveille. "Time to get up," said Pat. "And keep an eye out for the foxes. We should stay clear of them, so we don't end up with a fight on our hands."

Pat was a good leader. She told us to pack up, so we packed up. She checked her compass, and then turned to pick a landmark at 275 degrees. "Henry, you are in the left hunter spot. David, you are on the right."

Then she saw something move. Just a few steps ahead was a fallen tree. Mama Fox had dug out a space under the tree as their den. What Pat saw was three sets of eyes glued on our group and, undoubtedly, they were wondering what in the world these strange creatures are doing.

Pat maneuvered the group about ten yards to the left, repositioned us in our "formation" and said, not too loud, "Forward march!" Everyone now seemed to be focused on getting to the lake as soon as we can.

Later in the morning, maybe an hour before noon, we had just finished hiking over a small hill, and there, perhaps two hundred yards away, was the lake. We had lived in a space cruiser for four hundred years, so none of us had ever seen a lake before. This was definitely exciting. The estimate from our photos was the lake was about a quarter mile across. Our pace quickened, and our excitement level was rapidly increasing. Just a few minutes later, we were at the water's edge.

Henry and Richard, our successful hunters, were now interested in catching some fish for lunch. Pat reminded the group we needed to test the water, and then we also must test any fish we might catch. Melvin, our senior geologist, had been designated as tester. Earlier, Clyde Holdren, the senior Scientist, had put together a test kit for Melvin, just for this purpose.

Melvin found a convenient place at the water's edge to take his first sample. Looking like a chemist, Melvin took a look at the sample in his container, dropped in a small tablet, watched it carefully until it dissolved completely, then shook the container vigorously to thoroughly mix the solution. Melvin commented, "Hmmm. This appears to be OK. One down and two to go." Melvin walked to the left about twenty yards and repeated his test. He hollered to the group, "This test is OK. Only one more." With his third sample in hand, he rejoined the group. He did the test, then he explained to the group, "All three samples passed the test. The test we used is looking for the poisons that we might expect to find back on Earth. It is possible there are poisons on Alpha-1 that our tests will not find. Clyde recommends we should still be cautious. If anyone has a reaction to the water, such as stomach-ache, heart palpitations, or difficulty breathing, then we should boil the water before using.

Henry and Richard had found some articles on fishing. A lot of this was like trying to read Greek, because they had never even seen a fish. Let alone a lake. They brought some basic equipment, such as fishhooks and string. They cut a six-foot green tree branch to use for a fishing pole, tied a ten-foot length of string to the small end, and a fishhook at the end of the string. The instructions said grasshoppers make good bait. They spent nearly an hour looking for grasshoppers – none. And no bugs at all. The instructions also said worms make good bait. They remembered seeing some worms while digging a hole yesterday.

They started digging and found several worms right away. Henry fumbled with a worm for a while, then he finally stabbed the worm several times, and decided it wasn't going anywhere. He turned to the water and gently tossed the end of his line into the water. Where did it go? He figured the water was three or four feet deep, and now his bait was sitting on the bottom. The instructions said most fish are not bottom feeders. He brought in his line and tied a small piece of wood about a foot from the hook. He tossed the line back in. The wood was floating, so his bait and hook should now be about a foot or two from the surface.

Richard had been carefully watching Henry, and now he is ready to prepare his fishing pole. He moved a few feet away from Henry and tossed his line into the water. Richard turned to look at Henry. Both were grinning from ear to ear.

Richard asked, "What's the next step?"

Henry replied, "The article I read on fishing said patience helps,"

About that time, Henry's float disappeared. "Where did my float go?" Then the float was back. Then it was gone again.

Richard said, "That might be a fish!"

Henry raised his pole, and there was a foot-and-a-half-long fish. Henry maneuvered his pole so the fish would be away from the water, "Just in case he shakes off the hook."

A couple of minutes later, we heard Henry holler, "OMG, Look at those teeth! I don't believe freshwater fish are supposed to have teeth." Henry headed for the campsite so Melvin could check over his fish. Melvin studied the fish, then took out his pocketknife. He cut an opening on the side of Henry's fish, "Now we press one of these tablets to the flesh. If the fish is poisonous the area turns red. If it turns blue or green, it is OK." Eventually, they tested three fish and they were all OK. All of us were cautioned to be careful of their teeth.

Pat is not only the leader for this expedition. She is also the head cook. Now she seemed to be studying a problem, and then she explained, "I was planning on the rest of our venison for lunch today, and then maybe fish for supper. But these fish look so good, let's have them for lunch and then we can finish off the venison tonight. I think the meat will last long enough."

Henry and Richard did the basic cleaning of the fish. Norm and David dug a hole for the remains. This was a good-size hole, since we knew we might be here for several days.

Our three cooks, Melvin, Billy, and Pat were quick to start preparing a fire, and then laying out the utensils needed for the meal. When the fish was frying in the skillet, we moved up close, enjoying the aroma. We all agreed, the fish was delicious. Perhaps the best meal we ever had. The other part of the meal was an artificial veggie and fruit salad.

After we cleaned up the meal, Pat called us together to talk about exploring. “We are here to look around and take notes. I would like us to take a walk around the lake this afternoon. That should be about a mile, and a slow walk might take two or three hours. Let’s make sure our fire is out, and we can plan to leave in about ten minutes. By the way, we have some clouds up above. That might mean rain, so make sure you put your sleeping equipment and extra clothing inside a waterproof bag.”

We are walking on a path with the lake to our left. Pat has the lead, with Norm in a hunting position out to the right and slightly ahead of Pat. Richard is in a similar position, about fifty yards behind Norm.

For no particular reason we are spaced closer together than yesterday. Perhaps we could carry on conversations better. In any case, it a benefit to our mission. Nobody in the group had yet seen any sign of ants or spiders. We had counted eleven of those large owls, like the one that startled us the first night. We wondered, is it possible the owls are eating all the ants and spiders? This particular discussion turned on Melvin (our geology expert), who then gave us a couple of lessons on zoology (study of animals) and entomology (study of insects). He pointed out that an over-population of owls could eventually exterminate the entire population of insects.

While Melvin was talking, our two hunters reported seeing two more owls. The entire group was now interested in confirming that certain species of animals and insects existed, or not, on Alpha-1; or at least in the areas we were travelling through.

Back at our campsite, we had some free time before dinner. Several have waded into the shallow water. Someone asked Melvin if he thought it might be safe to go swimming. Melvin got excited, motioned for Pat to join the conversation, then said, "We have no idea what lies under the surface of this water. Paleontology is the study of extinct species in earlier periods, say five million to twenty-five million years ago, back on Earth. Scientists have found fossil remains of some terrible creatures. You have probably seen pictures of dinosaurs. These were huge creatures; some were vegetarians, but many were not. Imagine swimming and having to fight off an overgrown man-eating turtle. How about an early catfish, with shark-like teeth and large enough to swallow a human? And many other creatures that you can't even imagine. We do not have a clue how Alpha-1 compares to Earth on the evolution scale. Until we know a lot more, we should not go into the lake. Any lake. Oh yes, I don't think it is a good idea to wade more than a couple of feet from the shoreline."

Pat asked the group to join her to help plan tomorrow's events. "We have learned a lot here. What next? We could head back to our ship tomorrow morning, or we can take a look at another lake. That lake is about the same size as this one, and only five miles away. Due north."

Pat took a vote, and the team was largely in favor of visiting the other lake.

"OK, in about two hours we will have our second venison dinner. Then we should get a good night's rest, so we can get started right after sunrise."

Melvin, Billy, and Pat brought out the cooking and eating equipment, and examined everything to make sure it is clean and ready to go to work. Melvin did another water test and gave his ok.

Tom and Melvin decided to do some hunting. Pat told them, "You should plan to be back here about 4:30. And don't you guys go shooting each other. Or us." Grins all around.

Henry and Richard went fishing and brought Melvin and Norm with them. During this "school" session they learned how to rig their poles, dig for worms, and put the worms on the hooks. And they gave special instructions to avoid those sharp teeth. Before long each had caught two or three fish, and it was time to clean up for dinner.

The main course for dinner was venison, and side dishes included small portions of squirrel, fish, and artificial fruit salad. It was another successful meal, and everyone was in a good mood. They spent an hour or so telling funny stories and wondering what the folks back at the ship would be doing.

Eventually everyone gave up and crawled into their sleeping bags. It was a very dark night, and they were entertained by several night-owls, two barking foxes, a howling wolf, and something with a loud growl that was not-so-friendly sounding. Someone turned on a flashlight. Scanning the perimeter revealed reflecting eyes in almost every direction.

The light went off, and it was again a very dark night. Now it was very quiet, but the eight team members had some trouble going to sleep.

Day 3, Walk to Second Lake. Morning twilight seemed to come early. Everyone was awake, not hungry, and anxious to move on. They packed, and Pat said, "The map I looked at showed the second lake due north of the first lake, so we will take a heading of 360 degrees. Billy and I are in the hunter positions, and Tom is the temporary leader."

Tom cleared his throat, stuck out his chest and hollered, "On your mark, get set, let's go!" and off they went. All of a sudden, we are in a good mood. The most popular subject of discussion was the loud growl as they were trying to go to sleep last night.

They concluded it could have been from a wolf or a bear.

All of a sudden, Pat gave the signal for "all stop," and motioned for Billy to come join her. She had knelt in the grass while waiting for Billy and motioned for him to do the same.

Billy said, "What's up?"

Pat pointed up ahead, just a little to the right, and said, "What is that?"

Billy replied to Pat, "That is a raccoon - a very large raccoon. The articles I read said the adult raccoon normally does not exceed thirty pounds. That guy has to be closer to a hundred."

Pat asked, "Are they mean? Will this guy attack us if we miss?"

Billy replied, "Those short legs and awkward build usually make people think they should be clumsy and slow, while actually, they are quite fast, and very nasty in a fight. Their normal meals include smaller animals, such as mice, squirrels, and rabbits. I recommend we let him go and look for something we can handle."

Pat stood up and said to the group, "There is a very large raccoon up ahead. Maybe a hundred pounds. We don't need him, so let's leave him alone. But … if he charges us, we'll have to shoot him.

Tom, you can lead us on a path to circle around to the left."

Tom answered, "OK, gang. Follow me."

The very large raccoon must have seen us and decided to hide, or leave, because we never saw him again.

After a few more minutes, Billy must have come too close to a rabbit's lair. Billy was looking up ahead, and the rabbit jumped out from under a bush just a few feet from Billy. Billy was not prepared for this. The rabbit jumped in front of Billy and then was running away at his top speed. Billy brought his rifle to the ready position and squeezed the trigger. After a half-dozen shots, he remembered to aim, but by then the rabbit was gone.

Tom was laughing. As soon as he caught his breath, he said to Billy, "I'm sure glad the rabbit didn't run my way. That rifle might have been pointed at me."

Pat added to the levity, "If that rabbit had run toward Tom, you might have shot both of us."

Pat then told Tom it was time for a break. After everyone's nature visit, she said to the group," We need to talk about rifle safety. Who can give us a quick safety lecture?"

Richard stepped up and said, "A few days ago I got hooked on the subject of hunting and did some extra-curricular reading. Here are a few safety items that we should all remember. The first one is easy. When I am not a designated hunter, my safety should be on, and my rifle should be strapped to my backpack. When we are picked for hunting duty, keep the safety on until we start our walk. Here are the next safety items. Second, we should be holding the rifle in what is called the ready position. That is with both hands on the rifle and the rifle in front of us. Third, make sure the safety button is still in the on position. Fourth, your trigger finger should be on the trigger guard. Not on the trigger. Fifth, while walking, your vision should be primarily up ahead and back to around ninety degrees on the side away from the group. You will need to be looking all around your target area. Don't forget to look at the ground in front of you, so you don't stumble on a branch or something."

"When you have a target, or a possible target, bring the rifle up to a sighting position, and turn the safety button off. There are two sights. One at the far end of the barrel, and one just a few inches from your eye. You must place both sights on the target. Then, and not before, place your finger on the trigger. When you are ready to fire, and the sights are on the target, gently squeeze the trigger. Be careful not to jerk the rifle when you pull the trigger. When finished, set the safety back to on. If you were successful, then you should pick up your game, and I suppose you will want to show everyone what you shot. Any questions?"

There were no questions, but there were several teasing comments directed at Billy for his actions just a few minutes earlier. Tom came to Billy's rescue, "I think that's enough picking on Billy. We are all amateurs out here. If we had used our heads earlier, we should have asked Richard to give his lesson on rifle safety before we passed out the rifles. Come to think of it, some target practice might have been a big help too."

Pat took over, "Thank you Tom, and Richard. Good job. Both of you. Now, back to our business at hand. I believe we're ready to hit the road in a few minutes. Henry, you can have the left hunting position and David on the right. Any questions?" There were none. "OK, backpacks up." …short pause … " And let's go. Just follow me. I believe we should see the second lake in about two hours."

Two hours later, we found the lake. The scene was entirely different. The first lake was in the lowlands, and surrounded by deciduous, broadleaf trees. This lake is on the top of a hill, with primarily conifers, and many large boulders. Melvin said this might have been the location of a serious earthquake, or a volcanic eruption. We could not see a river or stream, so we assume the water is from a spring.

It was 10:30 and Pat said, "If you guys can go catch some fish, Billie and I will start a fire so we can have them for lunch." Melvin tested several water samples and reported, "This water is better than what we had at the other lake." Billy was to help Pat with the cooking. Henry, Norm, and David were the hunters, while Richard, Melvin, and Tom went fishing.

About ten minutes later, Melvin announced, "We're in luck, the fish tested OK."

At high noon, Pat blew her whistle and told the group to bring her their goodies for lunch. She decided to handle the red meat herself, which came from two squirrels and a rabbit. The fishermen brought in a dozen very nice fish. Billie was to start cleaning half of those, and Pat asked Richard if he knew a way to keep the others fresh so we can save them for dinner. Richard tied a string to a sturdy stick which he positioned between two boulders. The other end of the string was threaded through the gills of the six fish, with another stick tied at the far end so the fish could not sneak away. This was then placed back in the water.

An hour later, Pat blew her whistle again and announced, "Lunch time, come and get it." We did not have a real breakfast this morning; only artificial snacks, so everyone is hungry, and eager to dig in.

After lunch Pat called a brief meeting to plan our afternoon activities. We decided to leave our backpacks here, covered with waterproof bags, in case of rain. We were carrying rifles; in case we found a good hunting area – or encountered some large critters that might be pursuing us for lunch.

The time was two PM, and we figured the trip around the lake would take about three hours. It was a nice day and Melvin had to share his feelings on the lake, "Note that the rim of the lake is very uniform all the way around. At least as far as we can see. I probed the depth of the water, and found it drops off quickly, starting about five feet from the edge. All the gravel and boulders we see are probably from the inner part of this hill. We are definitely looking at the remains of a small volcano. And I would bet the source of the water is a spring in the lower part of the lake. If so, we should see some runoff somewhere." This was interesting to everyone, and they started looking for more evidence from the ancient volcano.

We are not Alone. A little further, Pat gave us something else to think about, "Hey gang, I need some help. Is that a footprint? You know, like a shoeprint or boot print? Or what? It is hard to tell with all the rocks and gravel."

Two people said, "It was nothing." The other four said, "Possibly a footprint."

Melvin said, "That's a footprint to me."

Norm said, “I wish I had those photographs from the orbit here so I could take a closer look. You know, Alpha-1 was an important project for NSEA. I wonder, is it possible they sent another space cruiser after we disappeared?”

Another in the group gave this response, “Hmmm … Now we have a LOT more to think about.”

Up ahead was a waterfall. At the bottom of the waterfall, we saw a smaller lake and a small stream that led off to someplace in the woods. This, then, is the runoff that Melvin said we would probably find.

There was no way to continue our walk in this direction.

It was three o’clock. Pat looked at her watch and then at Norm and Tom. She said, “We could go back and walk around the other way. If we hurry, we could be back at the waterfall around five-thirty, and then back to our campsite before dark.”

Norm and Tom agreed. Pat asked if anyone objected. There were no objections, and we were off – at a somewhat faster pace.

We were walking back to our campsite, where we started from, at four-fifteen.

At five-thirty we were nearing the waterfall. We found more footprints. Then we saw another path, leading to our left, away from the lake. To a lower level perhaps? To the water down there? Richard volunteered to walk part way on the path, while the rest of us remained at this position on the main path. At about fifty feet, Richard gave a series of hand signals that we eventually interpreted to mean, "You guys come over to this position, then I will walk further." During his second section, Richard made a sudden stop, turned around and rushed back to the group.

Pat said, "Richard, what's the matter, you look white as a ghost."

Richard started to talk, but he stuttered. He took a deep breath, then started over, "About twenty feet from where I stopped, there was a b-b-b-bridge. It was a rustic b-b-b-bridge, with Logs and rope. I didn't see any boards. And I did not want to go any further."

Norm said to the group, "My guess is we will find some people from another NSEA space cruiser. Probably friendly, and we will want to talk. And that is a problem. We will need several hours, and then we won't get back to our campsite before dark. And I don't want us trying to walk that path in the dark. I recommend we head back to our campsite, have a good nights' sleep and then be back here, say, mid-morning tomorrow. What do you think? Tom? Pat?"

They looked a little disappointed, but they agreed with Norm. The trip back to the campsite was quiet. Dinner was fish and some of our imitation veggies.

We sat around the fire for a while, talking about this and that. More than anything else, we wondered who they were and when they arrived. If they had been sent here to find the crew on our ship, they should have returned home long ago. Before long everyone was yawning, and we decided to call it a day.

Billy suddenly stood up and said, "Hey, wait a minute. Shouldn't we stay up for a while and make our plans for tomorrow?"

Pat replied, very calmly, "We are following the plan laid out by Norm, over there by the waterfall. That's enough detail for now. Good night."

Day 5. In the morning, the members of the group were eager to get started. The early risers caught a few fish, Pat made a fire, and breakfast was over at eight-thirty.

Pat stood up and said to the group, "OK, here are the plans for this morning. Let's leave our backpacks and sleeping gear here. Only two rifles. Richard and Melvin, are you OK with the rifles today?" They both nodded. "We don't want our new friends to think we are attacking them."

Pat poured some water on the fire, and said, "We leave in five minutes."

They took another look at the campsite to confirm the backpacks and sleeping gear were stacked and covered with the waterproof bags - and the fire was out. Our jovial bunch was on their way at eight-forty-five.

At ten AM they were close to the other path. Norm stopped the group and said, "We had a lot of conversation earlier, and now everyone is tongue-tied. I don't know what we are going to find, but I don't want them to think we are sneaking up on them. Let's keep the conversation up to a normal level. Everyone agreed, and the group returned to their style.

Meeting our neighbors. After a few moments we made the turn to the other path. Conversation continued, but at a somewhat lower level. We stopped at the bridge for a moment to examine it. Around the next curve, we saw a log cabin. Not large, maybe three or four rooms. Norm turned to Pat and gave her a signal to take over. Pat nodded slightly, thought for a few seconds, then said, "Hello there, anyone home? Hello, hello."

Then the door opened, and a man dressed in blue jeans, a long-sleeve shirt, and knee-length boots stepped out. Tall, good build, and a friendly face. He replied. "Hello there. Come on in. We were expecting visitors next week, and we thought they showed up early. They are from Our Town. Where are you guys from."

Pat turned to Norm, with a look that meant, "Your turn."

Norm was ready for this, and said, "We landed about twenty-five miles from here, about two weeks ago. My name is Norm, and I was the pilot; This is Tom, president of the board; This is Melvin, our head geologist, and this is Pat, leader of this team for exploring the neighborhood. "

Our new neighbor stood there with a shocked look. Then he smiled and said, "My name is Brad, and we have been here for almost four hundred years." Brad turned to the house, and said, "Linda, you need to come out here. We have visitors from Earth."

At the mention of "four hundred years," we knew this was going to be an interesting day. Norm was just as friendly as Brad, and said he thought it might be best to hear Brad and Linda's story first, and then we would give ours. Brad invited everyone to come inside, where all of us would be more comfortable. Brad and Linda had two children: Scooter, an eight-year-old girl, and Bobby, a six-year-old boy. We were a little crowded with a dozen people all total. Their dining table was a picnic table and adequate for eight. They pulled up some other chairs close enough to be part of the group. Linda had earlier made a small pot of coffee, and she poured half-portions for those who wanted coffee, and then started a new pot. She probably guessed that making coffee would be her major job for the next hour or two.

Brad started his story, "Our ship launched 398 years ago. We were aware that another ship had launched two years earlier and had disappeared. Our trip to Alpha-1 went exactly as planned. The long periods of strap-down for acceleration and deceleration were hard on us. We would have been bored silly if not for the books to read and movies to watch. I understand that orbiting the planet for several days was exciting, and so was the landing. We were the happiest people in the universe when the ship landed, after our nine-month flight."

"We commenced work right away on our mission and spent three or four months exploring the areas selected by our leaders. I am told that this period was actually very interesting. Our ancestors enjoyed what they were doing, and life was good. That is, life was good until we learned that we could not lift off. Now understand, there is a lot about this that I don't know, or don't understand, so you will probably want to talk to some of our technical people."

Norm just had to ask a question, "Are you going to tell us about the remaining 397 years? How did you handle the transition from visitor to resident?"

Brad answered, "More interesting stories. I am told that everyone was initially shocked. We had two or three suicides, but feelings started to change in a few days. We soon realized this may not be Earth, but it might be a reasonable substitute. We started building homes. We had already started looking for local plants so we could start growing crops, things like corn, wheat, grapes, apples, and so on. The soil here is quite fertile, and we seem to be getting the right amount of rain for our crops."

"We arranged some temporary houses in a small community which we named Our Town. We had assumed that someone would be here to rescue us. We were careful not to have children, so we would not add to the burden for an eventual flight back to Earth. After two years we changed our thinking and accepted that Alpha-1 is our home. We elected a mayor of Our Town and developed some rules to make life easier for us. We had already established a doctor's office, a food store, and a clothing store. When we hit the two-year milestone, we started getting serious about some other things; things like having babies and establishing some manufacturing capabilities. We were only two hundred and fifty people, so it was a very small town. Even today, we are still only five thousand."

"After four or five years, we built a computer factory, and established a computer network. A few years later, we turned one of our smaller buildings into a school, just for first through sixth grade. Twenty years later, we expanded the school. Then, more recently, just several years ago, we started building an automobile factory. No one here had any related experience, and we had to learn everything from scratch. Maybe another year or two."

Brad looked at Norm, "Norm, we are very eager to hear your story. Especially where you have been for the last four hundred years?"

Norm was on his third cup of coffee and ready to jump into this. He took another sip, "Brad, I am so impressed with what you people have done here. I believe you might find our story a little boring."

Norm started our story with a brief description of our normal lift off and early accelerations. "Then we had a problem as we moved to higher acceleration. We contacted NSEA and they told us to return to NSEA so they could fix it. Then we discovered we could not turn around, because there was a section of computer code missing, and we could not make changes to our scheduled flight path. Then, just a few days later, we lost all communication with NSEA, and we soon realized we were on our own. We finally determined the computers would still take us to Alpha-1, and then would give us control again for orbiting and landing. The bad news was that it would take us four hundred years for the trip."

"We had committed ourselves to completing our mission, and everything we did was designed to support that commitment. We accepted that our purpose now was to pass down to nineteen other generations that commitment and those instructions they would need in order to carry out the mission. Unlike your trip, Brad, ours was quite comfortable. Speed-wise, we were limited to just below light-speed, and the majority of the trip was at one G or zero G acceleration. We had several basketball courts, playing the regular version during times of one G and a new floating version during zero G. We had other games as well, such as computer golf. And yes, we read lots of books and watched lots of movies."

Norm described our education system, providing a full program from Pre-School to College levels; and yes, that was for nineteen generations. Norm also described our management arrangements, with the business of flying the ship in the hands of our four NSEA officers, and other business managed by an elected Board of Directors.

Norm continued, "Now that we are here, I believe we should offer to join your existing organization, and then we can assist you in whatever ways we can. But there is something else we need to consider. Given that both of our ships have problems, is it possible we can pull parts from one ship to fix the other?"

That question had been sitting quietly in everyone's head, just waiting to see who would be first to ask it.

Tom added this, "During these last four hundred years, we have worked hard to educate and train our personnel in many academic subjects that should be very helpful for chasing down the problems and developing plans for solving those problems. And, by the way, one of our people chose to pursue a master's degree in automotive engineering. That might be useful for your new automotive factory."

Brad smiled and answered, "Yes, that will certainly help."

Norm asked a second question, "Brad, do you know what went wrong; why you couldn't fly back?"

Brad quickly answered, "Honest, I don't have the slightest idea."

"How about any records, or a ship's diary, that might have some useful information?"

"I don't have an answer to that question either. How about if I call the mayor and ask him?"

Norm said, "You have telephones here?"

Brad answered, "Well, yes. Our system is similar to the old cell-phone system."

"Our strong suits are farming, computers, and communications. We have placed towers throughout our community, so computer access and phone communications are available to everyone. My property here is only two miles from town, and we have access to everything."

After another hour of conversation, Brad started preparing to call the mayor. "He has been in this position for several years, and everyone calls him Mr. Mayor."

Brad punched the buttons on his phone, "Hello Mr. Mayor, this is Brad … Yes, it is a nice day. Mr. Mayor, the reason I called is to share some interesting information with you. Are you sitting down?"

Mr. Mayor answered, "Well, yes, I am. Is this something scary? If it is, I want my assistant to be sitting here with me."

"Well, it is not scary, but it might be a good idea to have someone with you."

Brad heard Mr. Mayor call for his assistant. Finally, he said, "Ok, Judy is sitting here with me. What's up, Brad?"

"I have eight people sitting here in my house. Their space cruiser landed on Alpha-1 two weeks ago, about thirty miles from here. They have two hundred and twenty-five people on their ship, and a party of eight was dispatched in our direction to search for a drinkable water source. They found our lake yesterday afternoon and came across my cabin this morning."

"OMG. This is great. I want to meet them. Can you bring them to Our Town tomorrow morning? Say, around eleven, so we can talk for a bit, and then they can join us for lunch."

The Mayor's office was in the middle of town, and only two and a half miles from Brad's place, somewhat more than a one-hour walk. The schedule was agreed by all concerned. The Mayor had another comment, "And we will look forward to all the news updates you can give us."

Brad answered, "Well, no, they won't be able to do that. Their ship is the one that vanished two years before our ship was launched on our trip from Earth."

"Well, I will certainly look forward to hearing that story."

Lunch with our new Neighbors. Brad and Linda treated our team to lunch, which was primarily sandwiches with squirrel and rabbit meat between two slices of homemade bread. Corn on the cob was available as a side dish. None of our team had ever eaten corn on the cob. In fact, they had never eaten corn. The two kids took on the task of teaching us how to eat corn.

Suzie, the eight-year-old said, "I like to do it this way. I'm right-handed so I hold the cob with my left hand, between my thumb and these two fingers. Put the bottom end of the corncob on my plate. Pick up a sharp knife with my right hand, Start the blade at the top, like this, then push the blade down, like this, and you can see how the corn kernels fall in a nice pile on my plate. Move the cob around a little bit with my left hand and do it again. Then three more times and it is done. Oh yes, put a little bit of butter on the corn. It tastes better that way."

Bobbie, the six-year-old, was ready to show us his method, "This is the way the boys do it. I lay the cob down on my plate. Then I dip Dad's clean paintbrush in the soft butter, and then spread a lot of butter on the cob, rolling the cob back and forth a little bit, until the corn is soaked with the butter. Then I pick it up with both hands and eat it like this." The little guy munched through all those rows in less than a minute.

After lunch, Brad took us outside and showed us his farm. Well, not all of it, just the parts close to the house. He has a barn and four cows. None of us had ever seen a real live cow, but these cows were different from pictures we had seen. We asked Brad if these were real cows.

Brad was quick to answer, "Well, I really don't know. My Great-Great-Grandpa shot one like these while he was out hunting. He carried the remains back to the house then while he was separating the edible sections, he saw the heavy milk glands, and a light turned on. He eventually lassoed another one and brought it home. And then another. Pretty soon he had six of these. He kept them in the barnyard and started treating them like cows. In a few months, they had calves. He said the cows eventually developed a heavier build, and the milk glands also grew larger. Then he started giving cows to others in our community, and then everyone had cows. And here we are. They turned into cows. We know they are funny-looking cows, but they are cows. They give us plenty of milk. When they are too old to give milk, we butcher them for the meat."

While we were in the barn, we saw two bicycles. We asked Brad about these, and he said there is a small manufacturing plant, and they started making bicycles about a hundred years ago. Our Town was small enough to get by without them, but they were important for the farmers. All of the farms are within ten miles of town, and bicycles are very important to them. Brad showed us the cart he hooked up to his bike when he had produce to take to town or supplies to carry back to the house. "I started hearing rumors about a year ago that the factory was starting to design a motorized bike. They call it a motorbike. I have not seen one yet, and I haven't heard any new rumors on the subject."

Brad said he had ten acres growing various crops: corn, wheat, peas, and green beans. They did not bring any seeds from Earth, so they had to search for look-alikes growing naturally, and then try to make improvements. Brad recently found some grapevines and was experimenting in a smaller section behind the barn.

We asked Brad some financial questions, "Do you have a monetary system? You know, for buying and selling things."

He said, “We have the best financial system in the world – I mean universe. It is a trade system. Basically, we help each other. We don’t charge anything, although we might make a mental note: You owe me one. My Great-Great-Grandpa contacted five or six others to help build this log cabin. He signed a paper for each saying, Thanks for your help, and I will help you for a similar job.”

Meeting the Mayor. The next morning, we arrived at Brad and Linda’s place at nine-thirty. Linda declined to join for the visit to Our Town, saying she already had a full day planned for home-schooling. At Brad’s earlier suggestion, we brought all of our belongings and then placed what we did not need for today in the barn hayloft.

We started our walk a few minutes before ten. It was a nice day, and the path had been cleared so this is an easy trip.

We started seeing the near side of Our Town at ten-thirty. The style of the buildings reminded us of pictures we had seen of the early southwestern part of the States, back in the 1800s. And then we arrived at the Mayor’s office. “Come in. Please come in,” he said with a jovial tone. Then he turned to Brad for the introductions.

Brad did a good job of introducing everyone. Then the Mayor took over, “Let’s step into my conference room so you can meet some of my people."

When inside, Mister Mayor asked everyone to sit, "Have a seat and be comfortable. We will be passing out tea and coffee, and also some sweet rolls to take the edge off if needed. By the way, we will have a very nice lunch in about an hour.

Mister Mayor continued, "Those already in the room, when you arrived, are my two assistants, and the twelve elected members of the City Council. We can put names to everyone later. But now, I want to take a few minutes to point out the similarities of the events for the people from the two cruisers. They have both "disappeared" for four hundred years. We have all been deprived of our earlier lives and our expectation of our futures. In that regard, these two groups have had very similar experiences."

Mister Mayor paused to take a drink of water, then continued, "But not everything is similar, is it? For us who landed here 398 years ago, we have had the luxury of building new lives here on Alpha-1. Some of us have turned into farmers, others have become carpenters or home builders. Some have become metal smiths, making tools and other products. But we made a mistake. We spent too much time, too many years, waiting to be rescued. Our ancestors had fine educations. They were smart people. But look at us now. Most of us quit school at the eighth or ninth grade level. None of us has attempted to take a college course. Who would teach the course anyway?"

"And now, look at some of the differences. Look at the lives of those who were stuck in their space cruiser for 400 years. They established right away a governing system and an education system. These people have encouraged their children and grandchildren to continue in the footsteps of their seniors for 20 generations. The eight people you see sitting here are just as educated and just as bright as those who made up the original members of their ship."

"And now, just think about inviting these new arrivals to join us and help us improve our lives."

"Let's also think about another possibility. We have two inoperative space cruisers on our planet. Is it possible for the engineers and scientists to find a way to rebuild one of the ships to be flyable? But before we jump at that possibility, do we really want to do that? I believe what we should do for now is to start thinking about this. There may be some who would prefer to stay here. We could pick some representatives of both groups to put their heads together and help us make decisions on these matters."

"And now, let's step into our cafeteria where we have some nice choices for lunch. You can pick out what you want and bring your plates back here. When we are finished with lunch, we will ask our new visitors to tell us how they managed to survive living in a space cruiser for 400 years."

The cafeteria lunch was really a cafeteria dinner. There were several red meats, and they looked similar to what we were familiar with, but they tasted different: this was quickly becoming the best meal we ever had. One person said, "Cook the meat slowly, on a medium-hot fire. When nearly done, sprinkle on your favorite spices."

We asked, "What are these spices?" The answer was interesting, "Spend some time in the woods, and look for interesting plants. You can taste these yourself. If the taste is interesting, then you may find it even more interesting on your meat."

We also asked, "This bird meat was very good; what kind of bird is this?"

They are quite rare on Alpha-1. Someone caught a few about a hundred years ago. He brought them home, took good care of them and developed a flock of a hundred or so. These are egg-laying birds, similar to the chickens on Planet Earth. You can cook the eggs in a skillet, or you can let the hens take care of their eggs and you will find that each egg develops into a new baby chick."

None of our group knew what potatoes were. We were shown what the uncooked potatoes looked like, and we recognized them as part of the underground roots of some plants. "You can peel the potatoes, then boil them for fifteen minutes. Then you mash these with a large spoon, and you have mashed potatoes. There are other vegetables that are good to eat, although some might be too spicy. These include onions, radishes, cucumbers, cabbage, lettuce, spinach, and broccoli."

Wow. We have so much to learn.

These conversations had started, in most cases, in the cafeteria, and then continued at the dinner table. As a food conversation ended, there was a natural transition for questions to us about "How in the world did you manage to survive in the space cruiser for four hundred years?" The responses were varied, of course, but the great majority ended up with something like, "We decided from the very beginning that we wanted our children and grandchildren to commit themselves to meeting the goals of the original mission, as were defined by NSEA, and to continue in our educational fields. These conversations continued for an hour after we had finished eating.

Then the Mayor stood up and said, "Everyone, let's take a five-minute break while our kitchen crew cleans up our table. Then we will continue our discussions."

When we came back into the room, the Mayor, Norm, and Tom were in a huddle at one end of the table, making plans for the next part of our meeting. Everyone sat, they stopped talking, and Norm took the floor.

"Hello everyone. My name is Norman Link, and I was the pilot of the cruiser that landed three weeks ago. So, if you have any complaints about noise or nuisance, then I'm the guy to holler at." Everyone smiled and a few laughed. Norm continued, "We started our search for water over a week ago, and I am anxious to report that we found more than water," More laughs. "On a more serious note, we need to be on the road within a couple of hours, and I would like to give you some things to think about. While the Mayor, Tom, and I were talking, just a few minutes ago, we raised a question: What do we want to do? Should we join you here in Our Town, or should we start a new community?" There was a quick buzz of comments. Some heads nodded in the affirmative, and others shook heads in the negative. Norm continued, "I can see that you recognize the importance of this question. I suggest we all spend some time on that question. What do you want to do? What do we want to do? But we don't need to make those decisions today.

"And here is another subject to think about. We have two inoperative space cruisers. My group has several people with education and skills that will probably be helpful for analyzing the problems.

Hopefully, they can help us determine if one or the other is repairable.

"If we can repair one of the ships, then we will need to address another question. Do we want to fly back to Earth? Would we be welcome? Will we be able to communicate? If not, will we be considered friend or foe? We could be shot down before we can talk to anyone."

"These are some of the things that the three of us were talking about just a few minutes ago. Your Mayor said he would pick a few people to put heads together and spend more time on these questions. As soon as Tom and I arrive back at our cruiser we will start a group doing the same things.

"Now, tell me about your cruiser. Where is it? What kind of shape is it in? When should we send a team here to start looking it over?"

Mr. Mayor had a very apologetic look on his face and said, "Our ship is only a mile away. I am told it seemed to be OK when it landed, at least that is what appears in their notebooks. But then, two years later, we had a very minor earthquake – just tremors – and it toppled. Why don't we walk out there so you can have a look?"

Norm answered, "Yes, I am anxious to have that look. Our ship has a problem with the operations computer. I only need a brief time to see if there is any damage to your computer. Can we go there this afternoon?"

The group spent a few minutes packing up their belongings and visiting the restrooms. A half-hour later they were looking at the space cruiser, laying on its side. It was not an impressive sight, but there was no visible damage to the outside.

We were wondering about the inside. Norm broke the silence. He turned to Mr. Mayor and said, "Where are the doors? I want to go inside."

Mr. Mayor answered, "The doors are on the other side. We have an eight-foot ladder standing there to make it easier to enter. Tommy, one of my assistants, has been inside several times. I would like him to lead the way for your tour inside. Some areas are a little awkward and perhaps a bit treacherous. We only have four flashlights, so I believe we should limit the tour to four people. Only three of you and my assistant should do the tour. Norm, I assume you are going in. Who else will be joining you?"

Norm did not hesitate, "I will be taking Jerry Wyatt and Patty Andrews. Can we start now?"

When the ship stands properly on the ground, there are twenty-two floors. The primary means of going up or down is by elevator. But the elevator is not available, so we have to crawl through twenty-two flights of stairs. Norm said to the small group, "The stairs spiral around and are fastened to the outer wall of the ship. Yes, it is awkward, but I don't believe it is treacherous. When we come back for a more detailed analysis, we will have to ask the hobby-shop guys to make us some of those hats with the battery lights. Having both hands available will make for a much better trip."

A half-hour later they stepped into the operations center, which was the "home office" for the ship's officers. With the ship lying on its side, the six fixed seats were definitely not a place to sit. Pat opened the cabinet that held the computers. She and Norm took turns leaning in for a moment. Pat turned around with a smile on her face, "We are in luck. I don't see anything that appears to be out of place."

Norm snapped a few pictures, then said, "Now let's head on out of here." Climbing down the stairs only took half as long as climbing up.

We said goodbye to Mr. Mayor and his people. Brad led the way on a different path to his place. Linda had prepared a light dinner for us. Later, Brad took us to the hayloft and helped us arrange piles of hay so we would be more comfortable for our overnight.

Day 7. Heading for Home. In the morning, we had a brief visit with Brad, and then headed for our most recent campsite at the other end of the lake. We were there for a half-hour, just long enough to catch a few fish for lunch.

Pat then took command of the group and laid out our plans for the rest of the morning. “Back to our normal routine, I will have the lead, and I will be steering us on a course of due south. That is 180 degrees on my compass. Henry will have the hunting spot on the left, and David on the right. We have enough fish for lunch, so we don’t really need more food. Unless you see a nice deer. We can always take an extra break to clean up some venison. OK, let’s roll.”

Half an hour later, David took a shot at a deer. It was a long shot and he missed, but we gave him a brief cheer, just for trying. We really like venison. Another half-hour later, we made a brief stop. Norm and Richard took the hunter positions, and now we are off. An hour later, Tom and Melvin relieved the two hunters. Then, before long, everyone was complaining they were hungry, so we stopped for an early lunch. We quickly prepared a fire for cooking, while Henry and Norm took care of cleaning the fish. Pat and Billie started preparing to do the cooking.

Lunch was good, although somewhat less than some expected. After eating, Pat and Billie were the hunters for the next hour. Pat told the others, “I have been thinking about hunting, and now I just know I can bag one of those deer. You can plan on having venison for dinner tonight.”

The first opportunity appeared just ten minutes into this shift. Pat saw a deer, about fifty yards ahead.

She quickly brought up the rifle, turned off the safety, and was trying to place the sight on the deer, when it disappeared. “Darn,” was all she said. Tom was the temporary leader, and he said, “I was watching you, and I believe your actions were too slow. Think about the steps you must take. Get those organized in your head now, so you don’t have to work it about when the time comes.”

“That really makes sense. I will work on that.”

Within a few minutes, the same deer re-appeared, and seemed to be chewing on the grass. Pat raised one arm and then started maneuvering to a better spot. Tom raised his arm to signal stop to the group, then he turned back to watch Pat.

Pat now realized she was closer than before, and now was action time. She performed the early steps, all at once, and not really thinking about it at all. She placed the sights on the upper chest area and pulled the trigger. The deer went down, and the group cheered. Pat ran to the deer and the others were right behind her, crowding to look at the deer.

Billie said, “Pat, you need to give it a head shot to make sure it’s dead.”

Pat said, “I’m too nervous. I’d probably shoot myself in the foot. Would you do that for me?”

Billie stepped forward and took care of that.

Then Pat said, "I think this is a good time for a break. Who wants to help me clean this up?"

Henry and Richard proceeded to help Pat with carving the meat and putting the results in two game bags. Everyone seemed to be happy with the results. Venison for dinner tonight. Norm and David finished digging a hole and the others started dropping in the unusable parts. They covered up the hole, and this task was over.

Pat called the group together for a minute or two of planning, "Now it's time for Henry and David to be the hunters. We are still heading due south, and we should be back to our first lake in another hour or two. I think we should stop to catch a few fish for tomorrow's breakfast and lunch, and then we will head in our easterly direction. Actually 095 degrees. I suggest two more hours on foot, then we can enjoy our venison dinner. Tomorrow morning let's have a quick breakfast, and then we will need another four or five hours to reach our ship. Any questions?" There were none. Then, "OK, we will leave in five minutes."

Five minutes later, Pat took on her master sergeant poise, "Attention! Backpacks up! Forward March!"

The rest of the day went as planned. The stop at lake number one resulted in ten good-size fish, more than enough for tomorrow's breakfast and lunch.

While there, Pat made a call to Clyde Holdren, our senior scientist, to give him an estimate for our arrival tomorrow, and ended with, "Boy, do we have some exciting news for you."

The late afternoon stop was celebrated with a great venison dinner. Afterwards, we were feeling worn out and anxious to call it a day. But first, Pat, Tom, and Norm had a brief discussion about informing our shipmates of our exciting trip. Finally, that was done, and all were sound asleep just a few minutes after sunset.

Day 8. We were awake at sunrise, and eager for our fried fish breakfast. We had three hours on the road, a brief lunch stop, one more hour on the road, and then we saw the pointed end of our ship, standing proudly above the tallest trees.

Back Home. As we approached the front-yard area of our ship, it was apparent that Clyde had arranged for everyone to be outside for the welcome. The carpenters had done a good job. We saw a new platform, about five feet above the ground, with a dozen new chairs and two small tables. Clyde led the way for us to be "on-stage," and suggested we take our seats. Clyde had a brief introduction and invited us to tell everyone about the exciting news.

Pat started our story. She reminded everyone that our mission for this trip was to look for a water source. We found two lakes. The water was drinkable, and the fish were edible. She talked about animals for hunting, and then bragged about shooting her first deer the day before.

Norm talked about the other people we met, and how they got there. Our audience was stunned. Absolutely no one expected we would run across anyone on Alpha-1. He described the status of their space cruiser, "We don't know why they could not take off and fly back home. I would like our engineers to chase that down. Also, Pat and the other computer experts will take a look at the main computer and think about moving that equipment to our ship."

Tom talked about the people, "They made a terrible mistake. When they learned their ship was no longer working, they gave up. For two years they did nothing but wait to be rescued. That, apparently, turned into a habit and not much was accomplished during the first hundred years. The only education provided today is the equivalent of our first six or eight years. No high school, and no college. Farming was developed early, perhaps in the second generation. Housebuilding started about the same time. The early houses were more like crude huts. Slowly, more improvements were made, and they started building stone houses and log houses. Eventually, they built a sawmill, and then carpentry was re-invented."

“Their sciences and technologies started advancing in the second hundred years. Industries then included farming, carpentry, textiles, hunting, and fishing. In the 5th or 6th generations, they developed skills in electronics, metal fabrication, and others. Their population has grown to about 2,500 people. More than half of those live in a city named Our Town. The others live on their own farms, all within ten or fifteen miles of Our Town. Located in the center of town is a very impressive stone church, apparently copied after some of the impressive cathedrals back on Earth.”

“They have a simple government, with a leader called Mr. Mayor. The Mayor has a small staff, and everyone gets along just fine. They have a small manufacturing plant for computers, phones, and bicycles. Earlier, perhaps a hundred years ago, they built cable-based telephone and computer network systems. These appeared to be working, but maintenance requirements eventually encouraged them to convert to wireless. So, just twenty years ago they built a basic wireless system for both cellphone and computer network requirements. Within the next few days, they will be extending their wireless systems to include our area, all the way to SC12.”

“Mr. Mayor told me they started an effort just two years ago to develop an automobile manufacturing factory. It is interesting they do not have a currency or any financial systems, although they do have a practice of signing a note to reciprocate for specific tasks involving periods of help or assistance. That appeared to be similar to a bartering system.

“We have some things to think about. Let’s assume for a moment that we are here to stay. In that case, do we want to join our new neighbors and be part of their community? Or would we rather remain separate and start building our own community?”

“This afternoon, Tom and I will be sending out a basic plan for a series of meetings, and we will ask for comments back to us with any suggestions you may have, and also to indicate if you are interested in being a part of this committee. We want to spend some time reviewing our options and identify pros and cons of those options. We will eventually need to have a meeting with our new neighbors to share ideas and work out a plan to move toward our best possible future.”

There was good applause with Norm’s closing remarks. Someone in the audience asked, “Norm, are you taking questions?”

Norm answered, “Well, yes, but I don’t want to be tied up for another hour. What is your question?

“It sounded like both space cruisers are out of commission. Is it possible that we can pull parts from one to fix up the other?”

"Good question. We certainly won't know if that is possible until our engineers have a chance to examine the two ships. We will include this subject in our committee meetings. Any other questions?"

There were several more questions. They were good questions, and Norm was consistent in telling the crowd we would include these topics in our committee meetings. Finally, Norm said, "These are good questions, and I suppose you could keep me up here for a few more hours. I understand everyone probably has several questions, and I will ask you to include your questions in your response to the invite that Tom and I will be sending out this afternoon."

With that, Norm and the others from the traveling group walked up the outside stairs and then rode the elevator to their several levels with their bunkers. Norm invited Tom and Pat to meet in the operations office, "Let's have the meeting in about two hours. Right now, I need a nap."

The three met at 4 PM in Norm's office. They sent out a message two hours later with a brief outline of the primary topics that needed to be reviewed. They asked for comments and suggestions. The first meeting was scheduled for one PM the following day in the large conference room and would be broadcast live so everyone could participate.

Our Mobile Phone System. The technicians of Our Town are quite skilled with both computers and telephones. They established a computer factory about two hundred years after arrival on Alpha-1 and have more recently established communication links with the town's circuits. They have already opened a project to establish a temporary communication link for us with the telephone circuits in Our Town. They will need to position conversion devices at both ends. At the sending location, our phones will translate our voice to an electrical signal. The electrical signal is transmitted in all directions, and it should be good for 15 to 20 miles. For longer distances, we will require a relay station to receive and re-transmit the signal. The area where SC12 stands is about 22 or 23 miles, as the crow flies, and we are planning on a relay station at the midpoint and a final broadcast station close to SC12. For good reception the three stations should be above the tree-tops and we do not have the equipment to quickly create those tall towers. Our technicians said they already have a temporary fix for that. Some of the technicians in our "hobby shop" were preparing several large balloons that would be attached to a tall tree at each of the appropriate areas. Then the carpenters could go to work building the tall towers, which we figure might take another several weeks.

The new team includes Jim Anderson and Jerry Wyatt, both good engineers, and Pat, a skilled computer scientist who can show them the route, and who also wants to spend more time studying the computer in SC15, and perhaps prepare for hauling it back to our area.

This subject had taken nearly three hours of discussion, almost the entire time we had planned for this meeting. The remaining time was spent preparing a list of new topics to tackle at the next meeting. Our new #1 topic was to be answering the question, "Where do we want to live?" One choice was to live with them, with them in a leadership role. The other would be to settle somewhere else, under our own leadership. The meeting was adjourned.

The three members of the new project team spent several hours that evening making plans. We needed to carry about a hundred pounds of equipment and tools, and that seemed overwhelming. "That is entirely too much to carry; we can't carry that in a backpack." But then Pat suggested we carry the heavy stuff on a stretcher. If we were at the other end, we might have asked Farmer Brad White if we could borrow his two-wheel trailer.

Pat mentioned, "In our earlier trip we spent a considerable amount of time hunting or fishing for our food. It would be a faster trip if our kitchen could provide the food for us." The team members liked that idea. Pat called the kitchen and placed their order for four people, "Two breakfasts, one lunch, and one dinner. Venison if possible. And please leave the food on the first floor, tomorrow morning, between 5:30 and 6 AM."

At six in the morning, we were moving equipment, backpacks, and other equipment to the ground level. Already there were a dozen or so others that showed up to give us a good send-off. They also helped move some of our stuff. That was a big help, and also served to put everyone in a good mood. About six-fifteen, hugs and kisses were complete. Then we hoisted the backpacks and the stretcher, and we were on the way.

Where do we Want to Live? Later in the morning, a few minutes before nine, Norm and Tom were setting up the conference room. We had 35 attendees, and that just about filled all the chairs. Tom started the meeting by reminding everyone, "Our number one task for today is to identify the pros and cons concerning where we decide to live. Who wants to start?"

The first person said, "Well, I certainly don't want to move to Our Town. Oh, wait a minute, can we call it Their Town? If we decide to move there, then we can call it Our Town, Okay?"

Tom replied. " Well, yes, that is OK with me." He looked toward Norm. Norm nodded OK.

Tom turned back and said, "That is approved. Please continue."

"They have been here almost 400 years. Their technologies are not up to date. That's probably because they didn't establish an education system. A sixth-grade education certainly will not take you very far." He started to sit down, but Tom motioned for him to remain standing.

"Thank you for those comments. Can you think of any changes that might make living here more acceptable?"

"Oh good, now I see where this is going. First, put us in charge. We should provide one of our bright people to be the mayor. And maybe all the assistant mayors as well.

Electrical System. Another person asked a question, "Do they have electricity?"

Tom turned to Norm and stuck out his hand, indicating Norm would answer this.

Norm stood, "Yes, they have electricity. They were pretty quick to establish an electrical system. They started using a nearby waterfall as a power source for generating electricity late in the first generation. So, that tells us they were using their brains for that. Several generations later, they realized they needed more than the small waterfall. They searched through the internet that came along with their space cruiser and then learned how to use solar power."

Water System. Another voice, "I am wondering about the water system."

Norm was still standing, “They have had a decent water system for a long time. But more recently, they discovered their rusted iron pipes are not going to last much longer. They already have a project underway to replace iron with plastic in the older parts of town. That project will eventually include the whole town. Now, in addition to that, we must remember there are two kinds of water. Drinking water and wastewater. The wastewater systems are all ancient and are literally falling apart. They are currently working on plans to rebuild these. I don’t believe this is something we would want to take over, but if we can find ways to expedite these projects, that would probably be welcome. I believe all new construction should be using new technology.”

Telephone System. “Yesterday you mentioned they had a phone system. Can you tell us more about that?”

Norm was still standing, and he grinned. Tom was happy that Norm was able to handle these technical subjects so well, “The phone system is cellular and much of the design is copied from the cell-phone system back on Earth. We have four of our best engineers on the way today to help build a cellular line from their town to our current location. Without a phone system we would be stuck with a day and a half walking back and forth. This is not a difficult project. We will probably be able to test the new system in a couple of days. Clyde Holdren is the manager for our part of this project. He said he will call me with an update when things are close."

"Aren't the technologies for this beyond their capabilities?"

Norm answered, "Not really. Once you figure out what is required, it becomes easier."

Management. Tom raised his hand, "Norm, would you mind if I take us back to some non-technical issues?" Norm nodded and sat down.

Tom continued, "OK, let's talk about what changes it will take for us to be happy with settling in Their Town. I seem to recall that the gent who opened our meeting earlier had suggested we should have a large role in managing this effort. He strongly suggested we provide the mayor and perhaps all of the Assistant Mayors as well."

Another person commented, "I believe that is going too far. I might suppose the mayor is sort of their "hero," and they would not stand for that. If we can have two or three people in the assistant positions, I think that would give us a fair chance to prove we are smart enough, and then we could have more positions later."

Another comment, "I would suggest we not try to unseat the mayor. However, the Mayor has six assistants, each with different responsibilities. How about asking for three of the assistants to be ours?"

Norm turned to First Person, "How would you feel about that? Will that give us a big enough role to have a positive influence on the future of Our Town?"

First Person answered, "Well, yes, I suppose that would be OK. I think we will be able to eventually take over, don't you?"

"I don't think we really want to take over. We want to help make improvements to the management, and improvements to Our Town."

First person finally gave in, "OK, I like the way you described that. That is fine with me."

Education System. Tom said, "We are making progress. Now let's look at the subject of education. It was suggested that we take over the entire responsibility for education. We could start with a set of rules for attendance. This would emphasize that classes in Kindergarten through Twelve are required. The next step would be a schedule for staffing and then actually interviewing and preparing to hire the many teachers we will eventually need.

Tom described that the college-level courses would be optional, and staffing need not be addressed until necessary. We would need to explain to the parents and students the importance of advanced education. Tom said, "Personally, I don't believe this should be a problem. We already have a complete "education plan," as well as qualified personnel to put in the required positions. Any questions, or comments?"

There was a brief discussion, but no objections to preparing a plan for an education system. Tom said he would try to set up a meeting with the mayor to discuss a plan for establishing an education system, as well as preparing a new schoolhouse.

In the afternoon session, the primary discussions were about the "utility" systems (electricity, drinking water, wastewater, etc.), and preparing a list of the suggestions we would have.

Telephone System. The crew members working on the phone system in Our Town were making good progress. They picked a tall tree at the eastern edge of town and fastened a strong rope for the balloon and the necessary equipment. This was a slow process, requiring three to four hours. The communication transmissions would come from their computer factory and would then be received by the "balloon equipment" and re-transmitted in an easterly direction.

The team then moved to an area they estimated was half-way to our space cruiser. At that point, they hoisted another balloon with the same set of equipment as before. When finished, Clyde pulled out his phone and punched in Norm's number. Norm's phone rang and indicated there was an incoming call. Norm shouted to the group, "I have a call from

Clyde." Actually, all he received was clicks. No voice. Norm shouted "Hello" several times. But that was enough for Clyde. He received a few clicks and that told him Norm had received his call (his clicks). Clyde hollered "Thank you," and then signed off. Norm received two more clicks and understood.

The team had reached the goals for today and made camp where they were. Early the next morning, Pat led them down the path to our cruiser. They arrived around noon, and everyone had a good time with introductions and lunch, which was provided by the kitchen personnel. 21

Plans for Our Meeting with Mr. Mayor. Later that afternoon, the phone system was working. Tom punched in the Mayor's number and the Mayor answered on the first ring. They discussed the phone system for a few minutes, then Tom gave the Mayor a quick summary of what we would like to discuss in a meeting at the Mayor's place. The Mayor was delighted to hear that we were offering to take over the responsibilities for the education system. He said he would start his people thinking about the best location for the new buildings. Regarding our ideas for improvements to the utilities, he was surprised that we would be interested.

Tom's reply there was, "We are interested in any area that might be in need of improvements."

They discussed a tentative schedule for the meeting, "About a week from now," was the Mayor's offer.

Tom asked about reservations in the hotel. The Mayor laughed, "We don't have a hotel. I will check around for volunteers to take in overnight guests. That should not be a problem."

Later, Tom visited Norm, in his office, to discuss how to best handle the meeting with the mayor. They agreed to meet with the principal managers of the various parts of the education system in the morning. Then in the afternoon we called several engineers to help us with the utility systems.

Education. The morning meeting went quite well. Attending were managers from three areas: grades K - 6, grades 7 - 12, and college courses. These were excellent people, and very capable of describing the best way to achieve an excellent education system. Early in the meeting, they decided to look for another to join the meeting. This person will be a counselor, to help the high school students evaluate their aptitudes and decide which career fields might be best for them."

Engineering Support. The afternoon meeting was not so easy. Engineers generally feel that their ideal vocations would include designing and producing new or improved products. None of those present had taken any courses related to water systems providing drinking water to homes, stores, and factories; or systems to handle waste.

Norm made some progress with the following, "I would like for you to think about designing new or improved drinking water systems that would do the job better and last longer. And then you can start thinking about the waste systems. Remember, these systems have been there for many years, and we certainly don't want to keep installing new systems that are merely repeats of the old ones. Also, think about the processes and equipment we would use for removing and replacing the old systems."

The Big Meeting. Norm and Tom held several meetings with the others and eventually picked a small team of individuals to participate in our presentation to the Mayor. The team included three of our education professionals to discuss our approach to establishing and managing a full-service education program, two engineers to discuss engineering support for improving the utilities, and another two engineers to discuss expanding the automobile manufacturing plant they had started.

The total distance, by following the walking paths from SC12 to Our Town is twenty-five miles. Younger ones travelling without a heavy load to carry can make the total trip in eight or nine hours, but several of the older folks on this team will need twelve to fifteen hours. Norm and Tom started out planning for a total team of nine but were talked into including backups for the two older ones and also a nurse, bringing the total to twelve. Two of our engineers volunteered to do the cooking.

We were treated to a going-away brunch on the outside "stage" at nine in the morning and a very touching send-off at nine-thirty. After a few hours we were happy to see that the older members seemed to be keeping up with the rest of the team. We stopped for a late lunch at two and another for the evening at six. Tom phoned the Mayor with an update on our progress, and an estimated arrival in Our Town about two or three, tomorrow afternoon.

The engineers went to work right away, starting a campfire, and then starting to cook the venison for our evening meal. Conversations during and after dinner were focused on our presentation to the Mayor. Our sixteen-mile walk was certainly more exercise than we were used to, so the campsite became very quiet shortly after sunset.

The Next Day. Tom had set his alarm for six in the morning, and they were off on part two of the twenty-five-mile hike at seven. About two hours later, Norm phoned the number for Brad and Linda, "I just wanted to let you know we should be passing by your place in about four hours."

Brad asked, "Can we plan on you here for lunch?"

"Thank you, but no. We had an early-up this morning and just stopped a few minutes ago at the first lake. We should be starting lunch in another half-hour. I promise we will only be at your place for ten minutes. Be prepared for a few rush visits to the rest room."

Brad was very cooperative and said, "Norm, we'll be delighted for you and your troops to stop by. And by the way, Linda and I will be attending your presentation. In fact, we would like to join with you for the rest of your walk to Our Town."

We had reached the first lake at ten-thirty. Tom and Norm quickly rigged several fishing poles and gave some instructions to the newcomers. Then Tom announced, "And now you have to catch your lunch."

This went well with everyone, except the elderly reps from our education department were not about to pick up a worm, and certainly would never put a worm on a hook. It took less than a half hour for us to catch enough fish for lunch. Tom and Norm cleaned the fish, and the engineers did the cooking.

We buried the scraps, put out the fire with several doses of water, and were moving again just before one o'clock. At one-thirty we were approaching Lake #2. Norm led the way to show them the waterfall and then they took the side path for a brief stop at Brad and Linda's house.

The new ones on this trip – all but Tom and Norm – had never seen a real house before, so they wanted a minute to absorb this. One of them said, "Wow. I can't believe it. A real house!"

Norm was nearby and he replied, "In another half hour you will be looking at Our Town, with twenty-five-hundred people."

Brad and Linda were sitting on their deck, ready to join us for the short trip to Our Town. Brad jumped up, participated in the introductions, and shook hands with everyone. Then he said, "Mr. Mayor wants us to expedite, so we will have more time for our presentations this afternoon. I recommend we just keep moving. Will anyone need a pee break before we start?"

Welcome to Our Town. Five minutes later, we were on our way. Everyone was in a good mood and enjoying themselves. Just fifteen minutes later, we could see the taller, 3-story buildings, including two apartment buildings, an office building, and a church. Brad pointed out each of the important sights and led us directly to the Mayor's building.

Mr. Mayor then escorted us to a conference room, where we were served refreshments and told to be comfortable for a few minutes. Then the Mayor gave us an overview of what he hoped to accomplish this afternoon. Norm and Tom briefly described what we wanted, and the Mayor agreed that his objectives were almost identical. With that, we moved to the theater building, across the street, for our presentations. The Mayor had informed us we would have some residents attending, but as we approached, we had the feeling that the whole town would be here.

The theater was fairly new, with seats for 3,000. It appeared to be nearly full, so we figured all of the 2,500 citizens are here. The Mayor directed us to our seats on the stage. He and several of his assistants also had seats with us.

Mr. Mayor opened the presentation with a brief description of our schedule. He explained that our ship had departed Earth two years prior to theirs. "Then our forebears spent the next 400 years in their space cruiser." The Mayor called Norm to the podium and said he had a few minutes to tell us how they survived their 400 years.

Norm took the lead for this part of our presentation, He explained that our original Captain, Nick Logan, must have been an absolute genius, "He encouraged everyone to exercise when possible. He set up a board of directors to help prepare procedures, processes, and rules to manage our lifestyles. He encouraged our members to marry and start families. He also encouraged the Board to establish an education committee. This committee started out teaching the lower grades as our children needed at the time. Then the committee started adding additional grades, all the way through high school. And then they established additional courses for the college ages. This might sound a bit difficult, but we were fortunate that at least a fourth of our members were professional educators already."

Norm invited Tom to step to the podium and add some detail to our education organization. After some general comments, he introduced Ellen Davis, our Head of the Education Department. Ellen spent a few minutes describing her qualifications, with emphasis on her 30 years of experience. Tom then introduced Fred Jefferson, our head recruiter for new professors to teach the college-level courses. Fred described his 30-plus years of education experience, including 20 years teaching college courses, and three years' experience in recruiting. Both presentations were impressive.

Tom continued, "We encouraged the new generations to try to aim for academic courses and professional fields that were similar to those of their parents. This way, we would be better able to accomplish the missions that our ancestors had sworn to accomplish, starting 400 years ago, when we finally reached Alpha-1. Each year, as new students approached the later high school years, we encouraged them to pick an adult as a big brother or big sister to help them make the right choices."

Tom paused for a second to look for any comments from Norm. Norm gave a nod and Tom continued, “As our first-generation leaders died off, we felt alone, and many of us were afraid that those plans might start falling apart. It didn’t take long for us to realize that our future now rested in the hands of our present leaders. As we elected our new leaders, we had them repeat the same commitments as were agreed to by our first-generation leaders. ”Tom then turned to Norm and gave a signal for him to take over.

Norm moved to the podium. “It is important that you understand the reasoning behind educating our members throughout this 400-year-period. The leaders at NSEA had carefully picked a combination of skills appropriate for accomplishing the mission at hand. We felt it appropriate to encourage those in the following generations to pursue the same set of skills. We needed these skills so we would be capable of completing our mission. That mission had to do with studying the flora and fauna as well as the geology of Alpha-1, so we could provide a technical report to our bosses when we returned to Earth.”

“Now, there is an interesting question that we will have to answer, Will we return to Earth? The ship that brought you people to Alpha-1 has some damage, and I don’t believe I would want to trust it for a return flight. The ship that brought us here, just a few weeks ago, has a defective computer. We have a group of engineers looking into the possibility of transferring the computer from your ship to our ship.

This will be a slow process, so don't start making travel plans. Even if the computer exchange does solve the problem in that ship, we will then need to answer another question. Do we want to make the trip? Part of me thinks, yes, by any means. And the other part thinks, no way, I like it here. But we will come to that later."

"So, here we are, 225 well-educated engineers and scientists. Tom spent a few minutes describing to you our ability to develop a full-blown education system for the good citizens on Alpha-1. Now I would like to describe some other things we can do for you. The existing water system is literally falling apart. That includes both the drinking water and the sewage systems. We do not claim to be experts at either of these, but a few engineers and scientists will go far toward improving your systems."

"We were impressed when we learned you have a bicycle-manufacturing plant and efforts under way to start manufacturing motorbikes and automobiles. Obviously, a lot of work has gone into the planning for these. Again, participation by a few engineers and scientists can be a big help in this regard. We were thinking the new factory could also take on the job of manufacturing other equipment, such as lawn mowers, farm machinery, road, and highway equipment, and maybe even aircraft. And how about boats and ships? All of these should be under review by the Manufacturing Manager. Please understand, I am not proposing to take over any current functions or responsibilities. You might consider this a quality control effort, where we are always looking for ways to advance from good to better."

Tom joined Norm at the podium, "The education system we are offering will be in addition to what you currently have. Eventually, we would have 40 to 50 personnel involved. But this will be easy for us. Our current staff has people who, just two months ago, were heavily involved in our education system. These same people could go to work tomorrow, when we decide we have the need. I would suggest a new education building with 30 to 40 classroom-size rooms, and a large theater for presentations and ceremonies. I would suggest our head of the education department be a member of the Mayor's staff."

"We will also need housing for 90 families. Perhaps there could be an apartment house – that could possibly be transitioned later to a hotel or an office building. Perhaps there should be a new housing development, with maybe a hundred new homes. We certainly do not need large homes. Remember, we have been living for 400 years in our 12 by 12 bunkers. Should we have one of our people on the Mayor's staff to manage the housing department? And how about another person to manage streets and highways?"

Norm took over the presentation to discuss some technical areas. “I believe some of the most important help might be in the areas of drinking water and wastewater. I would suggest we establish a position on the Mayor’s staff to review the new plans before any further work begins. This person would be a senior engineer who would have access to our other qualified engineers and scientists who can assist as needed in the Engineering Department. Perhaps those from the Engineering Department could look at the electrical and communications systems as well.”

Mr. Mayor stood up, came to the podium, and stood there for a moment, as if still in deep thought. He turned to look at Norm and Tom and smiled. Then he turned to the audience and said, “I told you these guys are good.” There was a muffled response from the audience, and we all took this to mean they agreed with Mr. Mayor.

Mr. Mayor had been taking notes, and now he arranged some papers in front of him. “Our business arrangements have been in place for nearly a year. My Board of Directors includes myself, as Mayor and Head of the Board of Directors, with eight positions as Assistant Mayors. I will be keeping four of my current staff, and then I will have room for four Assistant Mayors from our new citizens. One of these will be Head of the Education Department and will eventually have a working staff of 40 to 50 well-experienced personnel. Another Assistant Mayor will be a Manufacturing Engineer, who will serve as an assistant to our current President of Manufacturing. We will need another Assistant Mayor to be head of our new Streets and Highways Department. This new department will also be responsible for layouts of new housing development areas, as they develop. The other Assistant Mayor will be head of the new Engineering Department, with responsibility to oversee our utilities, such as drinking water, sewage, electricity, and communications." Mr. Mayor then paused for a moment, obviously in deep thought.

"I was thinking about a Board meeting today or tomorrow, but that is too soon for Norm and Tom to make their selections for who will be doing what. Perhaps a tentative date of two weeks from today would be better. Norm and Tom, you can keep me posted on that. In the meantime, we will set up four rooms in the building across the street for temporary lodging. And, by the way, I want each of you to take a bicycle with you. I know some of the pathways are still overgrown, but the bikes are not heavy, and they should be of benefit to you for your trips. Right after you leave tomorrow, we are sending a crew to start cleaning up the pathways so, by the time you are ready for your next trip, you should be able to ride your bikes the complete trip. This should cut your travel time in half. And, in another month or two, we will be ready to change to motorbikes, and that will make the trip even faster."

Mr. Mayor then turned to the visitors, "Norm, Tom, and the team from SC12, I wish you to have a safe trip, and we look forward to seeing you again in two weeks." Then the Mayor faced the audience, held both of his hands out, made a stirring signal, and the audience started their applause. The Mayor motioned for them to stand and did his stirring motion again.

The applause continued until we reached the road outside.

The half-mile path from Our Town to Brad's house is easy, and all seven of us are on our bikes for most of that section. We took a few minutes for goodbyes with Brad and Linda, who supplied us with food for our evening meal and breakfast the following morning. After another quick pee stop at the house, we said our good-byes and were on our way.

Back to SC12. Our trip back to the space cruiser was a bit awkward. There are seven of us and we have ten bikes. We also have a bike cart, which is great for carrying our backpacks, other supplies, and the extra bikes. Most of the forest paths, however, are cluttered with undergrowth and we must walk our bikes almost half the way. We stop at the campsite near First Lake and prepare our evening meal of venison, corn on the cob, and some other fresh vegetables. The next morning, we are up early, knowing this will be a long day. Before leaving, we caught several nice-size fish to save for our lunch.

Later in the afternoon, still two hours out, we phoned our kitchen manager to have our dinner set up in the picnic area, "Seven dinners, three with fish and four with venison. Thank you."

We enjoyed our nice reception and are settled on the outdoor stage, combining our evening meal with answering everyone's questions. Norm and Tom give a good summary of our presentation in Our Town, and the warm response from Mr. Mayor. Someone commented, "That's great. Pretty soon we'll end up running the place."

Tom replied, "No, I don't expect there will be any chance of that happening. Not soon anyway. Remember, the board has nine seats, and we only have four of those. So, we are in the minority, and we don't have a chance of forcing anything they don't want. Also, eventually, we will be voting for about all of the positions. And remember, we might have about 200 people qualified to vote, but they have ten-times that.

Space Cruiser Repairs. Norm and Pat had hoped a simple exchange of computer programs would fix the problem. We now have a team of six computer experts looking into the possibilities of replacing the operational computer program of SC12 with that of SC15. They quickly determined the "easy" solution was not possible. Before very long in the project they found the notes from the engineers who investigated the problem back in the first generation, 400 years ago. The cause of the problem was obvious. A section of the program had been removed for updating, and then was not replaced in time for the launch. The computer appeared to operate properly, until someone tried to call for one of the functions that required the missing code.

Our first-generation engineers then tried typing in the missing section of code, but any "changes" were immediately rejected.

Now, with two space cruisers, they will look into replacing the entire computer. The computer had been carefully examined and it appeared to be in good shape. They had not been looking forward to carrying the computer the 20-mile distance, but using a bicycle wagon makes the trip not-so-bad after all. They decided the first step would be to uninstall the computer from SC12. That way, their "learning" is performed on the inoperative computer. The next step involved towing a bicycle wagon to SC15, very carefully removing the computer, and, of course, bringing it back to SC12.

After the replacement computer was fully installed into SC12, they examined every line of computer code of that computer with the typed version in the operations manual. They were in luck. The computer code included the part that was missing from our computer.

Pat had created the committee two months ago. The borrowed computer was now installed, and Pat had performed several "dry runs," thereby testing everything they could without a test flight. She now called a meeting of her committee to bring Norm up to date on their progress, and also to talk about next steps. Norm suggested Tom should be in the meeting, and there were no objections. The meeting went well, and Norm thanked the computer engineers for their hard work. Then someone asked, "Are we ready to go for a test drive?"

Norm smiled and said, "The ship might be ready, but we are not. Look out the windows there. We have four excellent workshops down there, and more than a hundred temporary homes. and more homes in progress. It will not be long before all the bunkers in our ship are empty. On top of that, Mr. Mayor has a crew of carpenters working on new homes in Our Town for us to move into. We will start moving our first families in about two weeks."

"In order to launch our ship, we would need to remove anything burnable out to at least a hundred yards. I think we will need at least a year before we can consider launching. I am certainly not saying we can't do it, but we will need to wait for at least a year."

There was a good bit of friendly discussion. After another fifteen minutes, Norm said to Pat, "Pat, how about doing me a favor. Put a note on your calendar to bring this subject up a year from now. Same group, one year from today. And we will probably want to invite Mr. Mayor, but we should leave that decision up to Tom and me.'

Telephone System. The electricians are doing a good job. We are only 20 miles away from Our Town, and everyone is happy with the cell phone system. The computer-savvy electricians made sure the phone system included the internet system. The only remaining work was to replace the balloons with permanent towers. Then we had an argument. Some wanted wood towers, and others wanted steel. Once everyone had a chance to say their piece, however, it was not a difficult decision. Iron ore was available at a convenient location, just a few miles east of Our Town. The site had been discovered two hundred years ago, and processing plants had been established about halfway between there and Our Town. Several blast furnaces were constructed and later upgraded and rebuilt several times, so their capabilities are quite respectable.

Plans were prepared, and the technicians went to work fabricating the parts for the first tower. Installing the first tower took almost a week. This was a slow process because they did not have a crane for lifting sections to the higher areas, and we had to settle for the old-fashioned methods using rope and pulleys.

Our engineers started preparing recommendations for the manufacturing plant to build a crane. That would make this job a lot easier, and it would also help when we get ready to build taller buildings in Our Town. Then we put in a request to build a truck to carry the crane and the parts for the towers.

When the first tower was ready, we pulled down the balloon and relocated the receivers and transmitters on the tower. Also installed on the tower was an array of solar devices and batteries, for power. The total project took only a month to accomplish, with three temporary outages lasting less than an hour for each.

New Homes in Our Town. After several weeks of planning, a large crew of carpenters and other related talents started working on our new homes. We originally asked for 100 new homes, but that was cut down to 70, including 50 with two bedrooms and 20 with three. A few of these had two baths. No garages, of course, but the lots were purposely large enough to allow for additions.

Things were moving slowly in the first few weeks, with only two completions per week. The pace, however, quickly increased to three and then four as good carpenters became better, and then great.

Tom gives the orders on moving. He personally visits each family before agreeing to assignments and schedule. The pre-move actions started two days before moving day, when they were provided with bicycles for everyone and two carts for hauling furniture and other personal goods. Tom also assigns two younger guys to help with handling the heavy stuff. Fortunately, the paths from SC12 to Our Town have been greatly improved, which reduced the average travel time to a little over two hours. The young ones post the times, some of which were very close to one hour.

Several weeks later, the manufacturing plant started releasing motor bikes. After a few broken bones and damaged motor bikes, the paths were posted with speed limits of 10 MPH.

When the overall plan was approved, it was apparent we no longer needed to build more temporary houses around SC12. Those families who were still in the ship's bunkers were encouraged to move to the temporary houses as they became empty.

The last family moved in about four months after construction first started. One of the families that moved in during the third week included a skilled wood worker and a hobbyist. He had created a hobby shop for making wood products for the families in SC12. After his move was complete, he set up his hobby shop in his house, and opened his business in a spare room. In his spare time, he added another room to the house. It looked a lot like a garage, but he preferred to call it his hobby-shop.

Education Department. Ellen Davis had been our own Education Department Head for the past three years, and she was looking forward to creating an education organization for Our Town. The current school building has four rooms, three for classes, and one for administration and storage.

Ellen did some quick math and estimated we should have about 32 students in each of the six active grades. Grades one through six will be taught in the schoolhouse. The teachers had already long ago worked out a temporary schedule, giving grades 1, 2, and 3 four hours in the morning and grades 4, 5, and 6 four hours in the afternoon. At the present time attending classes was not a requirement, although grades 1 through 6 were highly recommended.

Classes beyond sixth grade are currently optional, and the current class sizes vary between none and five students. In most cases, the parents of these children were teaching them at home.

Formal education in Our Town currently ended with sixth grade. That was coming to an end. Ellen presented a new plan to the Board. Effective the following Monday, attending the six grades of elementary school is now mandatory. Some of the parents argued, but Mr. Mayor supported the new rule. End of problem.

The next step was to interview the six current teachers. Ellen carefully pointed out that "Academic qualifications for grade-school teachers were not that important. That is, it is more important to have a complete understanding of how to present new ideas to your students, and to stress the importance of accepting these ideas as goals to achieve and improve on as their experience levels grow and mature.

Later in this meeting, Ellen told the current teachers, "I have described what I believe are the important aptitudes for grade-school teachers. If you feel you possess these aptitudes, then I will ask you to stay with us today, and we will talk about a trial period. If you do not feel qualified, then you are welcome to stand up and walk out now."

The meeting was a success. Three teachers left the meeting. The other three agreed to stay with us for a one-week trial period. They later proved to have a good understanding of what they were supposed to do, and they learned fast from our teachers. Three other positions were filled by experienced teachers from SC-12. Within a few days, all six of the newly approved grade-school teachers were teaching.

At the present time, attending classes for grades 7 through 12 was to remain optional, although Ellen made it clear these grades would eventually become mandatory.

Ellen sent a message to the parents of recent 6th-grade students, looking for volunteers to go back to school at the 7th or 8th grade classes. After several meetings, we identified forty-five of the Our Town residents who were interested in continuing their education. The new 7th and 8th grade classes then had about 25 students each. Later on, we were able to identify a few more Our Town youngsters for each of the higher grades. We estimated that in another 5 or 6 years the class sizes would finally stabilize.

Our initial plan for teaching college-level courses was to continue only with the teenagers from our ship, with an average count of about 3 to 5 students per year. As Our Town students passed the 12th grade level, they would then be considered for college-level courses. Our college recruiters and senior professors, however, came up with the idea of teaching some technician-level courses to those Our Town adults who had demonstrated talents in those directions. We estimated that the source for college professors would continue to come from our personnel for another 10 to 12 years.

More Schools. The plans were prepared in three steps. The first step was a new building for grades seven through twelve. As soon as that construction started, we added a second floor to the existing elementary school. This second step includes a classroom for pre-school children, rooms for administration and storage, and a larger meeting room. The third step will start after the first two are complete and will be a separate building for the college-level courses. After the three building locations are identified, plans will be started for athletic fields.

Medical Facilities. When we arrived on Alpha-1, we had two fully qualified MDs, with age differences of about 20 years. We also had two qualified nurses, and our senior nurse was taking courses leading to an MD certification. This amounted to four medical personnel to support our population of 225 personnel. Now, with a combined population of 2,275 total personnel, we need to expedite these numbers. We did some research to look for recommended medical professionals. We found several sources recommended not less than 30 physicians to support a population of 10,000 people, and we equated that to 7 or 8 physicians to support a population of 2500. We calculated that we could meet that goal in about 8 more years.

We expedited our schedule for educating the two nurses so one would achieve her MD in one year and the other in the following year. Already in the education program are several medical students. Our new estimates are for three total MDs in one year, five in three years, and eight in five years.

Manufacturing Plant. Located at the northeast edge of Our Town, the plant's history dates back about 150 years. The plant's purpose initially was manufacturing small tools, such as hammers, screwdrivers, hand saws, shovels, and a variety of farming tools. Fifty years later, bicycles became the most popular product. Electricity was eventually introduced and some of the tools became power tools. Motor bikes were introduced just a month or so after we landed and were the new "best thing ever" from the plant.

The plant managers were now looking into manufacturing a small automobile. They had no related experience, of course, but had found enough information in the ancient computer network to get started. We arranged a visit with the key plant managers and several of our engineers with a knack for automobile design and it was almost exciting. Their biggest problem was the lack of detailed drawings for parts that must fit together properly. A drawing of an automobile might be a good representation of how it should look, but it was no help for manufacturing those inner pieces.

Our engineers took a few minutes to sketch how the front-end steering mechanism works and how the various parts connected. The plant manager then asked the engineers if they would start designing the complete list of parts, "From the front bumper to the rear."

Each of the engineers was given a new motorbike, to expedite their travels back and forth. Back home, they took over one of our temporary buildings to set up several drawing boards for use to create the detailed drawings for a small automobile.

The blueprints were finished in a week. Production of a complete set of parts took a month. Assembly, painting, and final preparations took another month, and then the managers started driving around town to show off their new product.

Monetary System. The new system was a bit cumbersome at first, and some of the salary ranges did not seem to be fair. The Mayor's Board made a few adjustments, and those problems were taken care of. The Board also opened a City Bank downtown, and prepared flyers encouraging everyone to open an account and start a savings plan. The upper floors of this building were designed as suites and are available as rental offices or apartments.

One Year Later, Exploring. We continue to expand our presence on Alpha-1 and have completed initial exploration to nearly 100 miles in most directions from Our Town. This is a slow process, and we figure it might take several hundred more years by the time we have looked at everything.

We are careful to map the locations of lakes and other areas that might contain useful matter, such as drinking water, oil, iron, and other minerals. Several of our geologists took it on their own to help with these actions. In order to pin-point the locations of lakes and other items, they developed the ability to take pictures from several miles above the surface with a balloon. This was an awkward apparatus that required a long string in order to recover everything.

Soon after, several aeronautical engineers on the team designed a better method, and drones became a new product of the factory.

Telephone System. As interests developed in some of the key areas, we have been extending our telephone system. This is currently a difficult process. Positioning the towers requires heavy equipment for the tower, which requires a vehicle to take the equipment to the site, which requires more equipment to clear out the trees and undergrowth to construct a road. The aeronautical engineers said, “We need a helicopter,” and they are currently talking with the managers at the manufacturing plant.

The telephone system has been a success from the beginning. Minor improvements were made to reduce the weight of the receivers and transmitters. Electricians are currently looking for ways to extend the length of coverage so we can reduce the need for so many towers.

Monetary System. Mr. Mayor invited Tom to join him for the project to design the paper currency and the coins. Tom agreed with the Mayor’s ideas about matching the style of the American system back on Earth. Basic designs were finished, and several people were working on the art. The next step will be producing temporary samples and giving our citizens a chance to review and make comments.

Leaders for the other subtasks had little or nothing to show progress at any of the board meetings. When asked why, typical answers were, "I don't know what to do," and, "You can give this job to someone else." After two months of nothing, the Mayor relieved them of their assignments. The Mayor then took responsibility for the jobs himself and invited Tom to assist if he wanted. Tom did some research on our 400-year-old Internet for help. What little he was able to find was not very helpful. If developing a new economy, the advice was to pattern after the economy of a successful neighbor country and try to match prices. But we have no neighbors, so that was no help.

Tom spent some time re-reading Mr. Mayor's initial ideas for pricing. Tom started tinkering with the pricing, adjusting up or down according to his own whims. He specifically concentrated on salaries for the working people and prices for houses. Following several hours of this, he found he could not come up with salaries or prices that made more sense than Mr. Mayor's estimates. Eventually, Tom called for a group meeting of several who seemed to have an interest in finances and asked for their opinions. The group spent an hour or two discussing the material, although they were really just making sure they all understood the task. Eventually the group concluded that the overall plan, as prepared originally by the Mayor, was about as good as they could make it.

The next day Tom shared with the Mayor the results of our review, "Mr. Mayor, I had some pretty smart people review your plan with me, and I think you will be pleased to hear that they have no changes to your plan. They thought your plan was right on the mark. Now, how can I help with executing your plan?"

The Mayor said he would bring that up at the next board meeting.

The Mayor had not been idle while Tom and the others were doing their review. He called in one of the better architects to help lay out the plans for a new building that he said he wanted to be named City Bank. The front half of the building looked pretty much like they all thought banks were supposed to look. Half of the other half were offices for those who would be serving in management positions. The remainder housed the presses for both paper dollars and metal coins.

New Homes in Our Town. They started out with a team of six carpenters. They were good, and they became excellent. The team grew to ten, and they worked on two houses at a time. Seventy of these houses were for families from our ship. Ten of our families preferred apartments which they were provided in one of the new office buildings in the downtown area.

The overall plan called for eighty new homes. The Board approved completing the other ten homes, which they estimated would be taken by new couples following their marriages in the coming year. Our Town is growing.

Education Department. Ellen was doing a great job managing the Education Department. The new layout system provided ample space to teach grades K through 6, with one room available for storage and another for management (Ellen's office).

Medical Facilities. Our two doctors took over the remaining education requirements to advance one of our nurses to earn her MD certification. That is recently completed, and they are now continuing the education of our other two nurses. Procedures are currently being discussed privately, sometimes in the doctors' offices, or available classrooms, and sometimes in the doctors' homes.

We now have three doctors, and we expect that number will increase to five in two more years, and then to eight in another three years.

Manufacturing Plant. The plant started manufacturing motor bikes just a month or so after we landed. They were, and still are, very popular. They had intended to introduce small automobiles in the following year or two, but the need for other products caused them to change their priorities. We needed dump trucks and dirt movers for clearing the way for new roads, digging trenches for water pipes, and carrying our steel towers to expand our telephone system.

Automobiles are currently in the start-up stage, and several prototypes should be ready in another month or two. Optimists are saying, "A small-scale production line will soon be established, and we are expecting to produce fifty, maybe a hundred, cars a year." The realists are saying, "No, we will be lucky to produce one car every couple of months. We are still in the hand-made stages."

There is another group of engineers at the plant who are looking at airplanes and helicopters. They have already made several drones that have proven useful for carrying heavy tools. They have designs on the drawing board and expect to start both airplane and helicopter prototypes in a few more months. The first prototypes will actually be drones. If all goes well, they will move the aviation products to a new building.

Cruiser Repairs. Replacement of the operations computer was completed and tested. We have done everything we can do on the ground. In order to go further, we will need to launch and test the systems in full flight. Cleaning up the ground surrounding the location of the ship is 90 percent complete. All of the temporary homes have been dismantled and the reusable lumber has been moved to a safe distance. We can be ready in a day.

Norm is the ship captain, and our leader for the remaining actions that still need to be taken care of. Tom, head of our ship's board of directors, and Mr. Mayor are key in assisting Norm in working out any difficult decisions.

Whenever Norm is asked a "Should we" question, his response may depend on his mood for the day. For example, in a good mood he might answer, "Our ship had a problem in the operations computer. I believe we have fixed the problem, and now we need a flight test to confirm it is really fixed."

However, in another mood he might be inclined to give a more philosophical answer, such as, "Sometimes I think we should not fix it. If it is fixed, for what are we going to use it? One could say we need to complete the original mission, which has to do with our report about what is on Alpha-1. But 400 years have passed. I can't imagine anyone there still cares anything about Alpha-1. If they do, they are probably tired of waiting and have crossed it off their list. And think about this: Do we want to go back to Earth? People from the other ship have been here on Alpha-1 for 400 years. This is their home. And now it is our home as well. I really have no interest in going back. If we know the ship is fixed, we will always be trying to answer the question, should we or shouldn't we. Let's just stay here."

It did not take long for the entire population to learn of Norm's opinion on "Where is home." Norm has accepted Alpha-1 as his home, and he has no real interest in flying back to Earth. This had a general calming effect on the others, and no one would dare to bring up the subject again. With this general acceptance, we saw an increase in activities to move forward with improvements to our education, our lifestyles, and, surprisingly, the birth rate, increasing our population growth from two percent to three percent. Life was good; and getting better all the time.

Chapter 7 – Ten Years on Alpha-1

Our total population increased to 3000, and we expect the number will increase to 3300 in another five years. Our Town has 2500, another 500 are farm families, and another 250 are in a new town named West Village . The citizens of West Village elected Tom Kennedy as their Mayor. Mr. Mayor arranged to have himself promoted to Governor, during which he finally let everyone know his name is actually Maynard Jones. Tommy Smith was promoted from Mayor's Assistant to Mayor of Our Town.

Captain Norm Link always enjoyed looking at the stars. Stargazing from Alpha-1 was different than gazing from Earth. "Stars are stars you say. Well, that's not true." From Earth, star gazers of many eons ago had given names to many of the formations formed by the stars. Many of the names were taken from the Gods that were believed to exist at those times. Did they believe those Gods really existed, or were they figments of someone's imagination? We'll never know. Anyway, some of these groups have different shapes and deserve different names, and Norm decided to put some names to these interesting formations.

This was a pleasant time of the year. Not too cool, not too warm; Norm and his wife, Janice, enjoyed sitting outside in the evening with a glass of wine, while Norm made sketches of the more prominent formations. He had named two formations already, Fish and Bird. Why those names? Norm said, "Well, my reasons are simple. One looks like a fish and the other looks like a bird."

In order to give his sketches a common reference, he indicated the direction to North with an obvious arrow, pointing to the North. Norm had chosen to face North, or perhaps indicate a reference to North (at least magnetic north) for all his sketches.

Norm is a bright man, and he figured if he sat outside every evening, facing north, he would be able to watch the sky revolve completely over the course of one year. He decided that every evening would not be necessary, as long as he indicated the number of evenings between sightings. But how long was a year? He asked several people and heard estimates between 225 and 250 days. It was apparent that no one had thought of putting some effort into determining this number. Norm took on the task of determining the actual number.

More Visitors. After three or four evenings, he was looking specifically for a North star. If there is a North star, then it would appear to be stationary, with all of the other stars slowly revolving around it. This particular evening, he thought he saw a moving star. Just the opposite of what he was looking for. He turned to his wife and said, "Janice, I believe my eyes are playing tricks on me. Could you help me out here for a moment?" She answered in the affirmative, of course. Norm continued, "OK, sit here, and look to the North -- about this far above the horizon," He indicated an approximate level with his hands. "Now, see those two stars that are very close together? Focus on those for a moment." After another minute or two, he continued, "Does one of those stars appear to be moving?"

She answered quickly, "Yes, the one on the left is stationary and the other is moving – a little to the right." She was focused in on the star, then she stood up, "Norm, that star just changed direction. I believe it is moving toward us! What is this? What's going on? I'm going inside!"

Norm had dialed the Mayor's number, "Hello, this is Maynard."

Norm said, "There is something you need to see. Outside. Can you step out your front door and look to the North."

"OK, what is this, are you playing tricks on me?"

Norm answered, "This is no trick. You will be seeing what appears to be a very large star, and it is moving toward us,"

"OMG, what is this? What's going on?"

"I first saw this about a half-hour ago. It was about where I imagined our North star would be. I thought it was moving in a circle, then it started moving in our direction. Just the size of a star at first, and now it is much larger. This reminds me of ten years ago, when I was flying our cruiser around Alpha-1, looking for signs of civilization."

"What did you see?"

"I thought I saw something, a barn perhaps, on the first fly-over. Then I made a low pass in the morning, to verify. But I must have missed it, and then I convinced myself there was no civilization. I was wrong there. We found a safe place to land and then it was another month before we came across Our Town."

"So … what are you thinking? Is this another space cruiser?"

"That is my hope. And I also hope it is from the same organization that sent us up here. 400 years ago."

"So … What should we do?"

"Well, there is nothing we can do right now." Then they made their plans for tomorrow morning. Norm called Tom and invited him to join in the morning.

We have company. Norm, Tom, and Maynard met at the airport at 9:00 in the morning. Their helicopter pilot, David Andrews, was there an hour before. He had already filled the gas tank and completed his preflight checklist. All this was in hopes the new spaceship would show up in our area in the morning.

Norm said, “They probably saw something yesterday, from a high altitude, maybe 20 or 30 miles, and then came by last night and looked for the lights from Our Town. And now they should come from the same direction as last night, so they don’t miss us.”

An hour later, we saw a dot in the sky. A few minutes later we saw the spaceship. Almost overhead, the visitor started a zigzag routine. Norm explained, “That’s probably their way of saying hello. I sure hope they decide to land close by. I noticed their ship has a stubby set of wings. That gives them better maneuverability, and they should be able to bring it down a lot easier than our landing ten years ago.”

Norm turned to the pilot, “David, how about taking us up above the trees, and maybe we can watch where they land. Then we won’t have to spend a lot of time looking for them.”

David agreed and we were airborne just five minutes later. Not long after that, we saw the spaceship about twenty miles to the east, at a higher altitude, maybe five thousand feet. It appeared to be moving slowly, and then it stopped about ten miles away.

Norm said to David, “How about moving us closer? Maybe five miles.” David nodded OK and moved us in to five miles. We were all glued to the event, watching their ship slowly moving closer to the ground. Norm motioned to David, "Closer. Two miles." David nodded and moved us closer. We were all excited, and we felt honored to have such an excellent view of this dramatic event.

As the ship moved closer to the ground the thrust was kicking up a lot of dust and dirt. The exhaust fumes were mixing in and now the ship is almost hidden by the huge brown-gray cloud. A minute later all we see is the cloud. "I think they are down … Yes, they are down … See, the gray cloud is going away." Norm is re-living his landing ten years ago. He is excited. We are all excited.

David started moving us closer, "Norm, how close can we get?"

Norm answered, "I'm not sure. Just to be safe, let's stay at a half-mile. Can you put us in a slow circle?"

We circled a couple of times. Everything was finally settled, and now we could clearly see the ship. Norm tried to remember what frequencies he could use to talk to them, then he gave up. "In the old days, 400 years ago, everything was electronic. We used to have a frequency called Guard, and all aircraft, including spacecraft, were required to have that frequency tuned in for emergencies. I have been browsing Guard since we took off, but no luck. You know, 400 years have passed, and we have no idea how the newer communications equipment operates."

Norm continued, "I believe that just woke up my memory banks. The operations handbook had some very strict instructions regarding landings. NSEA told us for any landing – other than landings at NSEA – we are to remain in the ship for at least an hour after landing in order to allow the exhaust fumes and dust to settle. Ten years ago, I told our crew to stay in the ship until the following morning. So, that may be what these new folks will do."

David jumped in, "Well, that sounds like we might as well go home for now, and then come back tomorrow morning."

"Right on. Let's meet at the airport tomorrow morning, same time, to take another look." All heads nodded for that, and David headed for the airport.

The next morning, we gathered at the appointed time. As before, David was ready and waiting, and we were in the air a few minutes before nine. At just a few minutes after nine, we were within a hundred yards of the ship, starting a slow circle. Then we saw the ship's running lights flashing. Norm said, "They see us, and they are letting us know."

Norm told David to continue the circle. Half a minute later, Norm said, "OK, let's stop. This is good." He pointed to the side of the ship, "Look, right there, you can see the outline of the main entry doors, about thirty feet above the ground, and then down at ground level is a small door. I believe we should see someone come out of the lower door."

David interrupted, "Before we go any further, can we look for a place to park the bird?"

"Good idea," replied all three of the others. It didn't take long. There was an open area just a hundred yards away.

Norm asked David if he had a long rope. The answer was "Yes, and why do you ask?"

"Just in case we need to make a quick exit, you can take off and toss down the end of the rope so we can climb up.

David laughed, "Excuse me. I just thought about what you might look like, climbing up the rope. But you're the boss. I just do what I'm told."

Maynard laughed, "Norm, I don't believe you should try to climb the rope. Look at you. You must be two hundred and fifty pounds. And you're what, in your seventies?"

Tom turned to Norm, "Norm, I think they're right. If it comes to that, you and I can walk out, or run."

Norm gave them a firm look and said, "Okay, I yield to you comedians." Then, with a smile on his face, he said, "David, I believe you should stay with your bird. Keep your phone on. One of us will call you so you will know we're okay. If we're not okay, then instructions will follow."

After David shut down the helicopter, the others prepared to head for the visitor's ship. Norm turned to David, "Do you have a side arm? Just in case."

"Yes, I have my favorite rifle. It hangs inside the bird. Always ready for a fight."

"I really don't expect a fight. But this does make me feel better. You know, just in case."

Tom turned to Maynard, "If they ever come out, you can start with your Welcome to Alpha-1 speech. Then introduce Norm as the Ship Captain and Pilot."

As they approached, the two upper doors swung open, and they saw a crowd of people, waving their hands and yelling their hello. A minute or two later, the lower door opened, and one individual was standing there, motioning for us to follow him inside. Norm explained, in a low voice, "There is a narrow staircase inside, going up to the floor where the larger doors are open. It appears that our visitors are friendly."

Introduction. Inside, the Mayor took the lead for introductions, "Welcome to Alpha-1. Our forefathers arrived here a little over four hundred years ago. Our ship captain made what he thought was a good landing, but the ground was not solid. After a few months, we had a minor earthquake and the ship tumbled. The damage was enough we were stuck here. We made many attempts to contact headquarters but were not able to contact anyone." Maynard pointed to Norm and said, "This is Captain Norman Link, Pilot of SC12. Their ship launched from NSEA four hundred and twelve years ago and landed here just eleven years ago. Standing here, pointing to Tom, is the elected Mayor of our newest town, named Western Village. Again, we welcome you to Alpha-1. Now, may we ask about your intentions, please."

"Thank you for the welcome. My name is Larry Anderson, Captain and Pilot of SC253. On my right is Jim Womack, Operations Officer and my first assistant. Our purpose is the same as yours when you first launched, that is to investigate the natural resources, looking for elements that might be useful on Planet Earth."

Norm moved a few steps down the staircase and phoned, "Hey David, these are good people. They are from the same outfit as ours, so you can relax. Come on in and join us. We have been invited to stay for lunch."

David's reply was brief, "Sounds good. I'll be there in a few."

Larry continued, "I am anxious to hear more on the stories of both ships. I was told that SC12 launched in 2065, the ship had several problems and then communications were lost. NSEA assumed the ship was uncontrollable and it undoubtedly drifted off to eternity and everyone on board eventually died. Some NSEA heads rolled, partly because of the communication problem, and partly some revisions were left out of the computer code. Families of the lost individuals sued of course, and NSEA ended up on the edge of bankruptcy. We don't have the time right now, but please, I want to hear the rest of that story. NSEA recovered of course. Then, two years later, they launched SC15 with the same mission. I understand that all went well, and the ship made a successful landing. NSEA received a message with a landing report and then nothing after that. Perhaps the tumble of your ship upset something in the communication system."

"What you may not know is that NSEA sent another ship a year later with a search and rescue mission. That ship made it far enough to circle Alpha-1 two or three times. Then contact was lost, and they were never heard from again. I wonder, is it possible they landed some place, and neither of you know of the other. If they landed on the other side of your world, you would have no clue."

Larry now invited his guests for lunch, which would be in this same room on the main floor. In order for his staff to prepare lunch, Larry took his guests for a brief tour of the operations room, where he showed off his office and his tools for navigating the ship. After a few minutes, Norm said, "I'm sure there have been many changes, but, frankly, this looks the same as in my ship , which is now more than four hundred years old." Everyone laughed, and Larry said, "I believe it must be time for lunch."

After they were seated and starting to sample their lunch, Larry said, "I was just thinking about how the technologies have changed. Norm, your ship's flight time was 400 years. I know that was mainly due to problems with the ship. Maynard, I understand you made the trip as originally planned, so you were in your ship for seven or eight months, right?"

Maynard stood, "I wasn't there, but I understand it was eight months. And it seemed like it was 400 years." Everyone laughed. Maynard sat down.

Larry continued, "Well, our trip was only three weeks. The new acceleration systems are almost magic. We don't even use that term. We jump from zero to 20 times the speed of light in just a few minutes. We do two more jumps, and we are at our cruising speed. Twenty days later and we are there."

Norm stood up. "OMG, I spent sixty years in our ship, and I'm from our eighteenth generation. Can I ride back with you? Then maybe I'll want to do it all over again." Norm gave a big smile, and then, "I'm just kidding," and he sat down.

Maynard spoke up, "Larry, when you have some time available, I would like to take you to Our Town for a quick tour, and then to my offices for a celebration dinner. I wonder if we can pick a date now. Maybe tomorrow or the next day. What works for you?"

Larry answered, "I'll need a couple of hours in the morning to take care of some tasks. How about late in the morning, say around eleven o'clock?"

Maynard turned to Norm and Tom, looking for their opinions on a schedule. They both nodded okay. Then he turned to David, "How many seats do you have in the helicopter?

David answered, "We have a total of eight seats."

Maynard turned to Larry, "David can have us here at eleven in the morning. There will be four of us. Can you fill the other four seats?"

Larry said, "That's great. We will be ready and waiting at eleven in the morning. Then, maybe the next day, you can join me for a longer trip in our aircraft."

"You have a helicopter in here?" asked Norm, with a look of disbelief.

Larry answered, "Our crew for this trip is only forty-five people. The ship is about the same design as yours, and that gives us a lot of extra room to bring other useful tools. We also have a small combination bulldozer and forklift. Oh yes, we also have a small dump truck. By the way, we don't use gasoline for power. About two-hundred years ago, scientists discovered how to convert many types of atoms into nuclear power. One of the floors in the ship has an atom-blaster for creating atom-power. I can take a shovel full of dirt, feed it to our atom-blaster and it sorts through the dirt, saves the useful atoms and discards the remainder. That shovel full of dirt would provide more than enough energy to take this ship all the way back to Earth.

Norm looked skeptical, "And how does that work?"

"You ask how does it work? I don't know. I gained a basic engineering degree and then developed a strong desire to be a space pilot. And that, ladies and gentlemen, is what brought me – and my crew – to Alpha-1."

Maynard said, "That is amazing. This has been quite a day. Larry, we'll see you tomorrow morning,

The next morning. We met at ten-forty-five, took our regular seats in the helicopter and David started the engine. After a couple of minutes to warm up the bird, we were in the air. Five minutes later we were waving at Larry and his team who were standing on the main floor. David made a nice landing just a few yards away from the ship. David shut down the rotating blades and Maynard motioned for them to come on out.

Everyone finished shaking hands and locking seat belts.

Norm spoke, “We are taking a slight detour, so we can see what my ship looks like from the air. And, Larry, I wanted you to see the combination porch and staircase our clever people built.”

Tom added, “This provides easier access direct to the main floor and is a comfortable place to meet and greet. Just beyond our front yard we came across the home of Mr. and Mrs. Bear and their two cubs. We started giving them leftovers from our hunting trips, and they were tame after a few days.”

Maynard didn’t like us using up any of his share of the time, and he said, “We have a lot to see, so let’s head back to the north. Larry, we will fly over the farmhouse of one of my best friends and a very professional farmer. And then it’s only a mile.”

Fifteen minutes later, David landed at the helicopter pad on top of Maynard’s building. Maynard had arranged for his staff to bring in coffee, tea, and snacks while he acquainted us with the schedule for today, “We will have an hour in the conference room to get acquainted, an hour in the dining room for a scrumptious meal, and then two hours in the auditorium.”

Maynard started with a brief introduction, and then Larry had a turn to talk about their mission and discuss how we can all assist the SC253 crew in meeting the goals of their mission.

When Larry started his story, we all groaned with envy when he mentioned their twenty-day flight. We had pretty much forgotten about our mission, which was to document the resources available on Alpha-1 and prepare a detailed report for NSEA. It was an important mission, but, after four hundred years, who remembered? The next part of Larry's talk was very interesting. He said his buzz-craft was currently circling Alpha-1 at a very high speed. This was on autopilot and under control of their computer system, with high-speed cameras taking pictures of the entire planet."

Larry continued, "When I return to my ship this afternoon, I will be meeting with some specialists to start studying the results. Then we will pick some of the more interesting areas to start visiting. Norm, would you be available to join us for the next couple of days?"

"Larry, I would be happy to help. How about if I bring along my head geologist? He is really good."

Larry looked like he just won a prize. "Yes. By all means. I have two good engineers, and twenty young and inexperienced geologists, so I can really use that kind of help. That will be a nice addition. I can pick up the two of you at the airport. Will eight in the morning work for you?"

"That's a yes for me, and I'll have to give you a tentative yes for Melvin. I can call you later today to confirm. Are you picking us up in your helicopter?"

"No, that's too slow. I'll be flying my buzz-craft. We won't reach the speed of light, but we might hit two miles per second now and then." Norm thought for a moment, then had a shocked look on his face. Then he gave it a thumbs-up.

Larry continued, "My mission is to document the types of natural resources available on Alpha-1. We will start studying the pictures this evening, and then we will start visiting some of the identified areas tomorrow. These visits may take as much as a week. My ship has some heavy equipment so we can dig down and load examples to take back to NSEA headquarters. Altogether, we should be finished and ready for our return flight about two weeks from now."

"Maynard, before we leave, we will have some gifts for you." Maynard leaned forward, to listen better. "We have some heavy equipment that we can leave with you. This includes a small combination bulldozer and fork-lift, and also a small dump truck. Remember, I mentioned we don't use gasoline for power? We will also leave for you our atom-blaster. That part of the equipment is very sensitive and difficult to keep in operation. If that is too much bother, you can change from atom to gas with the flip of a switch."

"And here is my final wrap-up for today. I have met only a few citizens from Alpha-1, and I wanted to say I enjoyed meeting you, and I look forward to meeting more citizens, and working with you. I certainly appreciate your offer to help us meet the requirements of our mission."

Larry turned to his host, "Maynard, in particular, I appreciate your hospitality, and I look forward to working with you these coming days." Larry turned, shook hands with Maynard and sat down.

Maynard felt a lot better after Larry's kind words. "Larry, I know I will enjoy getting to know you better in the next two weeks. If you have any new needs, just let me know." Maynard checked his phone for the time. Actually, it was just a signal that he was about to bring the meeting to a close. "This is amazing. We accomplished much more than I had expected, and we finished sooner than I expected. This is probably one of the most interesting meetings I have ever attended. Thank you, Larry, for the participation of your team. Norm and Tom, many thanks to you and your team. And, to the citizens of Alpha-1, thank you for your participation. And I hope you will be available if and when you might be called upon to assist Larry and his crew if they need."

With several grandiose motions, Maynard said to his audience, "Ladies and gentlemen, this meeting is over."

A few minutes later, we gathered in Maynard's conference room for refreshments. We were there for maybe fifteen minutes. New technologies or schedules were not discussed. Small talk only. Maynard led us to the roof and pointed to the helo pad, where David was cranking up the engine for a brief warmup.

In the air, Norm had a question he was anxious to hear the answer to. "Larry, you mentioned some heavy equipment you will be leaving for us. Will you also leave your buzz-craft?" Larry laughed, "And you haven't even seen it. The buzz-craft is not government material. That is my personal vehicle. Sort of like a scooter that I brought along for personal use. It has been inspected by the NSEA engineers and it is approved for my use in support of government business. No, when I leave, it leaves with me."

David parked his bird near the front of SC253 and turned off the engine.

Larry put his hand on Norm's arm and pulled him to the side. "Norm, it's only four o'clock. Would you and your geologist be available to work with us for a couple of hours this evening? You can help us pick out the more interesting areas to start visiting tomorrow morning. The truth of the matter is, I only have myself and a bunch of young geologists. My other smart guys are all engineers, and they won't be much help for this. Having you guys will double my team."

"Yes, I can help. But I need to talk to Melvin." Norm pulled out his phone and made the call. Melvin was more than happy to help. He sounded excited, and anxious to get started.

Larry said, "Norm, I'm going to bring out my buzz-craft so you and I can go pick up Melvin when he's ready."

David carried Tom and Pat back to the airport in his helicopter.

Larry took Norm inside SC253 and then they walked up to the first floor. Larry pushed 2 on the elevator button. The elevator door opened at the second floor, and we were looking at Larry's buzz-craft. Larry motioned to a marked area in front of the elevator and said, "Stand there, and watch this."

Larry pushed another button and several sections of the floor under the buzz-craft folded away, leaving the buzz-craft hanging by a cable in mid-air. Larry pushed yet another button and the buzz-craft was lowered to the first floor. "And now we can walk back down to the first floor."

By the time they were there, the cable had detached from the buzz-craft; then it disappeared to the second floor, and the floor sections moved back to their places.

Norm started looking at the buzz-craft, and talking to himself, "Four seats, with a steering wheel and foot pedals in front of the driver's seat – make that the pilot's seat. Some gauges, switches, a bunch of buttons and a small screen in the console. A plexiglass cover, providing full visibility to all. Behind the two back seats is a covered section, probably for the engine. Behind that are three heavy duty exhausts or thrust ports.

Larry told Norm, "Climb in and I'll point out a few things."

Larry gave Norm a quick review, then pushed a button and a small pair of wings appeared. "Our minimum flying speed is 105 mph. We have to get up to flying speed right away, so this is like a catapult from an aircraft carrier as you might have seen in an old movie. Fasten your seat belt. Now, hang on to anything that is loose and lean back in your seat. Here we go!"

Norm's response, "Wow! Zero to 120 mph in a fraction of a second.

Larry said, "That's a little fast to fly around town."

Larry pushed another button and the wings expanded to four times the original, then he pulled back on the throttle. "Now our minimum flying speed is 85, but I don't like to go below 95. Now, that should be Melvin's house about eight or nine houses ahead of us, right?"

Norm answered, "That's right. We're close … three… two ,,, "

"Now I push this button, and we can hover … like a helicopter."

Norm saw the wings swing in and tuck themselves away. They disappeared into that space beneath our feet. Then he heard a loud buzzing sound from the thrust ports. He commented to himself, "Hmmm … buzz-craft."

We were about 100 feet above the ground. Speed 35. Slowing. Descending. We're there. Nice landing.

Larry opened the top, and said to Norm, "It's time to go to work. Your first task is to introduce me to Melvin."

Larry knocked on the door, then stepped aside so Norm was closer to the door. Melvin opened the door, "Hi Norm. Say, that's a sporty-looking runabout. What is that?"

Norm answered, "First, let me introduce Larry Anderson, pilot of SC253. He and his crew will be here for a couple of weeks."

"Larry, this is Melvin Sherwood, our senior geologist."

Larry took over the conversation for a few minutes. "This here is called a buzz-craft. They were introduced twenty years ago and are quite popular on planet Earth. "Melvin, take one of the back seats there and we'll zip back to the cruiser."

Larry took off in a very careful fashion. Lift-off was slow and straight up. We saw the wings extending and then Larry leveled off at three hundred feet and carefully picked a path that would keep us clear of most of the houses. We had a good time waving to the neighbors. Out in the open, Larry pushed the throttle forward and said, "You should hang on to any loose items, then lean back in your seats. I'll give you a brief demonstration."

We felt the Gs take effect. We were shoved into the seat cushions, and we could not move our arms or legs. Norm wished he had his G-suit. Then the acceleration diminished, and in another minute, we were at a much slower speed, headed for Larry's cruiser. Larry, with a huge smile on his face, asked, "Did you guys enjoy that?"

Norm answered first, "I am really impressed with the flexibility. The only thing I have flown was our cruiser, and that took a long time to change speed or direction. How fast were we going back there?"

We are doing one-twenty right now. Earlier we topped at about five hundred."

Melvin asked, "Five hundred what?"

"That's miles per hour."

Norm and Melvin looked at each other. Mouths open, saying together, “WOW!”

We slowed down and approached the second-floor opening. Larry took the buzz-craft to a halt in front of the opening. Then the buzz-craft did a one-eighty degree turn and we backed into the opening. At the right time the power turned off and all was quiet. Motioning for us to stay in our seats for a moment, Larry stepped out and continued his tour of a few more buttons and switches. Then we stepped out and Larry pointed out a few more.

“Understand what we may be doing in the next several days. The ship has twenty-two levels, with the lower eight as storage space for tools and machinery. Two more levels are open space for specimens we might want to take back to NSEA. We have two levels for recreation. They contain exercise equipment, dart boards, several ping-pong tables, and a small basketball court. Two levels have several meeting rooms with large working tables. Six levels are for bunkers, and two levels for cruiser operations. Larry said he wanted us here for a couple of hours today so we could work and better understand.

Larry took us to Floor 14, where he introduced us to John Collins, his young geologist who would be part of our four-person team for initial inspections. There are two floors with work rooms. Floor 14 is currently configured with six large tables. The tables appeared to have glass surfaces, which are actually view-screens. Each table had a "matching" surface also on the wall next to the table. Larry started explaining, "Today we are working on the northern hemisphere. This film was taken yesterday, with the buzz-craft on a high-speed assignment to take pictures on a route to cover ninety east-west strips around the circumference of the planet. Each strip covers one degree of latitude, which is about 55 miles in width. Actually, we have about ten miles of overlap, so we don't miss anything. The lower latitude (from the equator to 30 degrees north) strips are longer than the others, so we have three tables working on this group. The next group covers the area between 30 degrees north to 60 degrees north, with a total width of 1,650 miles. We have two tables assigned for this group. The third group covers everything above the 60-degree latitude line. These strips are very short, so we only need one table for this area."

"Each table has five people assigned, all of whom have recent BS degrees in geology. They can work on the visuals in either the table viewer or the wall viewer. Workers will start by identifying areas of interest by marking the perimeter and giving it a reference number. If one wants a close-up view of a sector, he takes control of one viewer while the others can continue their tasks on the other viewer. The user will tap "Notes" for adding more information such as location of rock formations, good area for rock samples and fossils, or perhaps mark the distance away from an interesting landmark, or the distance across a river or lake, or describe other physical properties of the area."

"As areas are finished, I want the four of us to identify areas of more interest, so we can take a closer look. Obviously, we can't look at everything. My bosses have allowed me two weeks. I expect we will need one or two days for initial visits of the northern half. Then we send out several teams for four or five days of gathering samples. And then we use a similar process for the southern hemisphere."

John had been separated from our small group for a few minutes to talk with the teams working on the lower strips. They had finished the first four strips, and additional groups should be finished at about fifteen-minute intervals.

Larry and John set up the inspection table with the first four strips. Everything is electronic, so we don't have any paper to keep track of. The screen is touch-sensitive, and keying in specific text is replaced with checking the boxes for the items that apply to the many areas we examined.

We finished the first group in thirty minutes. Then we finished our remaining groups in twenty minutes each. John spoke up, "This is fun."

Norm did some quick math, "At this rate we should finish the first thirty degrees of latitude in two hours and forty minutes."

Larry spoke up, "Norm and Melvin, would you prefer to quit early this evening? Then we could finish first thing in the morning"

Norm answered, "Personally, I think this part is more important than examining the areas on paper. I would rather we keep going. Then we can start with the next task in the morning. How about you, Melvyn?"

Melvyn did a slow stretch, then answered, "I agree with Norm. Let's keep going. But I need to call my wife. Otherwise, she'll be worried."

They finished just before eight. Larry took Norm and Melvyn home, and they stepped out of the buzz-craft in front of Melvyn's house. Larry said, "I'll pick you guys up in the morning at seven-thirty. OK?"

Norm answered, "That's fine. See you in the morning."

Alice, Melvyn's wife, opened the door, "I'm glad your home. I was lonesome." The three of them chatted for several minutes.

Norman said to Alice, "I do believe Melvyn will enjoy these projects we are working on."

She replied, "Well, good. I was getting tired of watching this bored man walking around the house, looking for something to do." But, Norm, please see that he comes back home every evening."

The remaining walk to Norm's house was less than five minutes. Janice was standing there, holding the door open, "I heard the buzz and then I watched you land. How is it going?"

Norm stepped inside and gave her an overview of what they were doing. And then, "The final goal is for Larry and his crew to prepare a report on the natural resources available here, and to collect some samples of anything interesting. Now, think back to 400 years ago. You know that our cruiser lost communication and was believed to be lost in space. The next ship was launched two years later with the mission to do what our ship did not. That ship lost communication during the landing, and then the ship was damaged, and they were stuck here forever. And now, today, I heard Larry mention another ship was launched a year after that. That ship had a search and rescue mission, but communication was lost while orbiting or landing. One of Larry's missions is to search the surface for any clues about what happened to that ship. Was that a crash landing and everyone perished? Or was the landing successful and their 20th generation grandchildren are still here?"

Norm sipped from the glass of wine handed to him, then continued, "Larry said he prefers this part of his mission is not public knowledge, so please don't mention it to your friends. I am sure we will eventually be able to discuss, so save it for then."

Searching the Planet – First Day. The next morning, Larry took Norm and Melvin to the 13th floor where John was waiting for them. This room had three large tables, but only one was activated. Appearing on the glass are two strips, one for a section with the south end at the equator and the north end at 30 degrees, and the other strip with the same dimensions and adjacent to the west of the first strip. Each strip covers about 1500 miles.

Larry explained, "I would like to make one run today, on the eastern section. We will be going from south to north, and I marked six areas where I want to stop and look around. Two of the areas are generally flat land with a lot of trees. Two other areas are on the sides of mountain tops. I am assuming that the flat land and mountain top areas will have different resources. The other two are riverside areas.

Larry said, "I want you to see some things that I found very interesting. The three of you know that one of my missions is to find out what happened to SC16. So, I am always looking for signs of a crash landing. I would guess a crash landing would cause quite an explosion, and then would probably be followed by a fire. But what would that look like 400 years later? It might be overgrown by trees, and nothing is visible." Larry pointed to several areas that could possibly be the location of a crash landing.

Then Larry moved the pictures to a different area. “On the other hand, what should we be looking for if they successfully landed and are now citizens of Alpha-1, like the crews from SC12 and SC15. Here is a possibility. Look at this spot. It is completely covered with trees. Now let’s zoom in and take a closer look. Look at this small brown section that appears to be something on the ground between these trees. Is it possible these are from the roofs of buildings? This might be exactly what we are looking for. I want to take a closer look at this area. That will be our first stop. If there is nothing of interest there, then we scratch it off my list.”

Five minutes later, they are in the air. We were being shoved back in our seats, so we knew Larry was in a hurry. Norm asked, “How far away is this place?”

Larry answered quickly, “2,042 miles. That will take us about thirty minutes We’re presently loping along at 4,000 MPH.”

Thirty minutes later, they were at the site. Larry put the buzz-craft into a comfortable descent. “There. We can see the brown color. Going lower, we still see brown.” He moved to an area that had no trees. The brown was still there.

He lowered to about ten feet. “Look down there. We are in hover mode, and you can see the ripples in the water. The brown we saw was nothing more than muddy water. Well, I can scratch that one off my list. Over there, about fifty miles to the north is another place we need to look at. We can also get out and stretch our legs a bit."

Ten minutes later, Larry landed, and we all jumped out. Melvin had insisted last night that we check this out. Melvin had a question for Larry, "Is it ok for me to take samples? A few rocks is all I need."

"You can take a couple of handfuls, no more. If you need more, then we will return with the trailer. The air-trailer fastens on the back of the buzz-craft. Handy as can be."

Melvin pointed to several small ponds, "These small ponds were caused by a volcanic eruption. Just like the second pond we found ten years ago when we were searching for a water source. And the next day we visited with Mr. Mayor in Our Town."

Norm laughed, "That was an interesting day."

Melvin took several pictures and shoved a few rocks into his pockets, then said to Larry, "I have what we need for here."

In the buzz-craft, Larry said, "We have six easy ones. I just need to take some close-up pictures, and then we move on to more exciting areas.".

One of the "easy" ones looked very interesting. There was an attractive river, fast-moving and loaded with huge boulders. The terrain on one side of the river was nearly a hundred feet higher than the other side, with cliffs reaching down to the water. After buzzing slowly for a mile or so, Melvin said, "Can we stop for a few? I'd really like to see inside that cave."

Larry positioned his buzz-craft right in front of the large cave opening, which was halfway down the cliff. In hover mode, the buzz-craft was steady enough and Melvin could step onto the flat area in front of the cave. Larry asked Melvin why he wanted to look inside.

Melvin answered, "Back on Mother Earth, there are many ancient caves that have some interesting prehistoric artwork, some of which are more than 20,000 years old. I only need to be inside for a couple of minutes. And the only equipment I need are my flashlight and camera."

Larry answered, "You also need a sidearm."

"And why would I want that?"

"In case you're not alone in there. Some of those prehistoric people might still be in there. Sorry, that's a joke. More important, how about bears or wolves? Or tigers?" Larry pulled out a handgun and handed it to Melvin.

Melvin put the handgun in his pocket and stepped out. Then he cautiously moved into the cave. He turned on his flashlight. Two minutes later, the cave must have had a turn. All we could see was some light reflections from the walls.

Larry heard something, "Melvin, I heard a noise. Are you all right?"

"Yeah, I just stumbled. Clumsy me."

A minute or two later, we heard Melvin yell, "Yippee! I found some paintings on the walls. I will be famous. I mean, we will be famous. I'll take a few pictures, then I'm coming out. There are some bones on the floor. Probably bears or wolves. It is too spooky in here."

Melvin finally stepped out of the cave. He stumbled again and looked like he might fall over the edge. He recovered quickly, "Sorry about that. The light was blinding me, but I'm OK now."

He handed his phone to Larry and said, "The large painting is a view from the mouth of the cave, looking downstream with the terrain on the left side higher than the right. We will have to bring some experts to look at this. They can do some tests to determine how old this is."

Larry was quiet and seemed to be focused on the pictures. Then he pointed something out to Norm, and they both started laughing.

Melvin was offended, "And just what do you find so funny about my pictures?"

Larry spoke up first, "Well, you can cancel the call for the anthropologists. Your artists left some of their tools. Several cans of spray paint on the floor. You didn't see these when you were in there?"

"No, I didn't see any paint cans. But that could be what I stumbled on. Come to think of it, I did hear something metallic. Well, I guess I just wasted everyone's time."

Larry answered, "No, no. Not at all. This evidence indicates there must be more people on Alpha-1 than we thought. My report to NSEA will include this information. Now we will want to set up a serious search plan to locate them. Let's head back to the cruiser so we can go back through our pictures."

On the trip back to his cruiser, Larry explained, "My instructions say if we find evidence of SC16, that takes priority over everything else." Norm and Melvin quickly said they wanted to help.

Back in his cruiser, Larry assembled the team, "We will do this part in two steps. Our first step is to review the pictures and determine the specific areas we want to search. Then step 2 will be searching those areas. For now, we are only looking at the area within five miles of the cave. Let's do step 1 this afternoon, and then tomorrow morning, we can take everyone out to the site and start step 2."

“Now, please form into four groups, with six of you in each group. Each group should be looking at one of the quadrants that we will refer to as North, South, East, and West. We will use the cave as the center of the overall circle. Now pay attention to the orientation. If we are standing outside the cave, we can see a small river at the bottom of the steep wall. Then face the direction that the river flows and that is roughly East. The quadrant that we are looking at on our right is the East quadrant. The next quadrant on the right is South. On the other side of the river are areas West and North.

SC253 was sent to Alpha-1 with a primary mission of exploring the geologic capabilities. Their primary staff included 22 recent graduates of geology education programs and they all were proud of their new geology degrees. Larry had asked for his head geologist to be John Collins, with seven years of geology. They had flown a number of missions and enjoyed working together. As luck would have it, Norm’s staff on SC12 included Melvin Sherwood, with thirty years of geology experience.

Melvin had developed an interest in geology during his college days, and then he taught the subject up until time to land on Alpha-1. John and Melvin were both given the title of Senior Geologists, and they promised to carry on a friendly competition for the assignments.

"Your task for now is to examine the close-up pictures and identify any areas that we need to examine more closely. For those areas that are completely covered by trees, we will have to walk through to see if there is any evidence of habitation. For the open areas, you might be able to visually determine from the pictures if there are any signs of homes or campsites. If so, we want you to visit these sites, then let us know right away what you have found."

"Then tomorrow morning I will take everyone out to the site. We will attach the sidecar to the buzz-craft. The sidecar has room for six people, so we will be making four trips. The site is 2,000 miles away, and these trips will be about half an hour each way."

Someone commented, "Wow! Looks like we'll be traveling fast."

"Right. The buzz-craft has a maximum speed of more than double that, but we'll stay safe. For the next couple of hours, I want you to walk through all the areas you marked in your quadrants. We should all start from a spot near the front of the cave."

"Let's talk for a minute about what we're looking for. If we are lucky, we could come across a small town tomorrow. Their ship was on a search and rescue mission, and they had a small crew, about thirty or forty people. So, if we come across a village of maybe seven or eight hundred, then we know we hit the jackpot. Keep in mind, they may have given up waiting for a rescue. We don't want anyone to start thinking we are here to ruin their lives and force them to do something they don't want. In any case, remember to be friendly."

A hand went up, "I don't understand. What do you mean."

"Well, after four hundred years they may have accepted this as home. We need them to understand we are not here to force them to pack up and return to Earth. NSEA will be delighted to know that their crew survived the trip and the 400 years since. NSEA will treat them as heroes. And between you and me, NSEA will probably give them anything they want."

Another hand was up," What happens if we find nothing in the first five miles?"

Larry smiled, "Then the first phase is finished. The second phase will have a different approach. We'll probably program the buzz-craft to move our search beyond five miles with higher resolution low-altitude pictures. The pictures will be continuously transmitted back to our ship, so we can watch the search as a near-live video; or we could review the new sets of pictures periodically. Anytime we see something interesting, we can bring in the buzz-craft and take a closer look."

Working uniforms are handed out, and include trousers and long-sleeve shirts, all in a light-gray color. Ballcaps are provided in several colors: yellow for North, orange for East, blue for South, and green for West. Each team member also received plastic canteens. Larry described more that we would receive in the morning, “One-half of the team members will receive hunting knives and 30-caliber rifles. The other half will carry hatchets and handguns. These items are included not because we expect to need these, but rather so we will be ready for the unexpected.”

“You will be provided with lunch in the morning, Nothing fancy. You have your pick between PBJ (peanut butter and jelly), roast beef, and chicken sandwiches. Also, in the lunch boxes will be buttered corn on the cob, a half-dozen cookies, and a drink that is similar to Gatorade.”

Larry continued, “I will also be handing out new phones in the morning. They are for communication among us and with those back in the ship. I will be the overall coordinator, so if you find something interesting, give me a call. Then we go from there.”

After another three hours of studying the films, reviewing the many comments by the team, and highlighting the more important areas to look at in the morning, it was time to call it a day. A few minutes later, Larry was taking Norm and Melvin home. “Thanks, guys, for your excellent help today. My crew will be okay someday, but right now they are just beginners. I need you to keep your eyes and ears open tomorrow and watch for decisions that don't make sense.

Their replies to Larry confirmed they understood, and they would do exactly what he wanted. Melvin was to have the lead with the East group, the first group to start their search. The second group was South, with Norm as the lead. The third was West, with Larry's assistant, John Collins, as lead. Larry had the North group.

As they landed in front of Melvin's house, Larry said, "Thanks again, and I'll be picking you up at seven in the morning. "

Back at SC253, in the morning, Larry prepared his first report to NSEA to be brief and still cover all of the important events. He picked a title that would be easily recognizable: Report #1 from Alpha-1. He described our looking in the cave as the most important event of the day. We took this to mean there are people in the area, perhaps survivors from SC16, and maybe not far from this location. We returned to SC253 to make our plans for performing the search tomorrow.

Day 2. Everyone was excited about today's event. Norm and Melvin were standing out front when they heard the buzz-craft approaching and starting to come down to their level. Norm commented, "I was expecting it would look clumsy with the side car attached. But that looks pretty neat."

Another minute, and Larry said, “Good morning,” and opened the canopy so Norm and Melvin could step inside. Larry and John were in the front seats, Norm and Melvin had the two back seats. The other six were seated in a single file arrangement in the side car. Everyone appeared to be in a good mood.

The buzz-craft left the ground and Larry pushed the throttle for some speed as soon as the altimeter showed they were at a thousand feet. Larry commented, “Basically. it functions the same as before, but our additional weight is nearly three-thousand pounds, so it is a bit clumsy. I must start maneuvers a second or two earlier for turns and landings, especially landings. It took a couple of hard landings to teach me just how important that is.”

Half an hour later, they slowed down and then landed just over the cave. As soon as the others were out, Larry was anxious to go. Just before he started the buzz-craft, he spoke briefly. “Melvin, this group is yours. I’ll be back in an hour with Norm’s group. After that, we’ll be ready to start the last two. Give me a call anytime you find something interesting. ‘Bye for now.”

Melvin’s directions were brief. At a casual pace they could easily do one mile in an hour. That’s five hours for the trip. But with our eyes on the ground, looking for something, a better estimate might be an hour and a half for a mile. At the beginning of our walk, we will be bunched together. Melvin had some specific instructions.

"This is a good time to take out your compasses. Let's all look at our compasses and then face North. Note that the reading on your compass is 360 degrees, or zero degrees. The center of our quadrant is 360. The width of the quadrant is 90 degrees. Our left border is in the northwest direction, 45 degrees, so read that as 315 on your compass. The right-side border is at 045 degrees. So, the individual on our far-left side should have a course of 320, just 5 degrees from the border, The next person should be inside of number one by a hundred to two hundred yards."

"On the other border, number seven will stay on a course of 040, and number six about 100 to 200 yards inside of number seven. Number four will be at the center of the fan, and stay on a course of 000, or 360 if you prefer. Number three will be 100 to 200 yards to the left of number four, and number five the same distance to the right of number four. Any questions?"

There were no questions, "Good, then let's get started." Off they went, at a somewhat faster pace than planned, but that seemed to be OK, as long as they were so close together.

The day before, while comparing notes on the North quadrant, there was some discussion about several items that appeared to deserve a close look.

The most interesting item was a large lake, perhaps two miles across, positioned about 2 to 4 miles from the starting point, and between 020 and 030 degrees. It was agreed that number five would walk the shoreline on the west side and number six on the east side. Upon finishing the lake perimeter, they would then return to their original assignments.

One hour into today's walk, Number Five had moved over about a quarter mile, so the lake should be appearing soon. A few minutes later, he saw Number Six, still about five hundred yards to his right. He waved, and Number Six waved back. Ten minutes later, they joined and shook hands. Just short of another hour, they were at the lake shoreline.

Number Five commented to Six, "This looks like a well-worn path. Do you suppose there are people who use this path frequently?"

Six replied, "I suppose that's possible, but the path could be used by deer and other animals. I don't see any footprints. Let's keep each other advised as we continue our search."

Half an hour later, Six came to a group of trees just a few feet back from the shoreline. "Looks like a neat place to take a rest, maybe a short nap," he thought to himself "If I were fishing, I would stand right here and toss my line out in the water. Swing my pole back a little, and make sure I don't tangle with those branches, - like someone else did !" Right over his head was a long piece of string, with a bobber and hook at one end. Someone had tangled with the branches and could not shake it free, so he just cut the line and left it there.

Six hollered across the water to brag about his findings. They were about a quarter mile apart. Five returned Six's call, "That's great. It means we have humans involved here. Or nearby. You're getting all the fun stuff. I guess I'll have to work a little harder. I will look for footprints."

"Good luck on your side. See you at the North end."

Five was there first, and therefore the discoverer of a thatched-roof hut. Primitive, yes, but a good way to stay out of the rain. The hut was built around four sturdy posts, from four-inch logs. The posts are six feet tall in the front and four feet in the back. Connections are still firm. And they appear to be made with some sort of home-made rope. The roof is neatly covered with one-inch branches and something that looks like bamboo. Laying on top of that was a collection of large leaves to direct the rainwater to the back of the hut. By the time Six arrived, Five had taken several photos, and jotted down a half-page of notes.

Six commented on the remains of a campfire, about five feet in front of the hut, "This was probably for cooking; and perhaps also for heat in the cooler months."

Six said, "Right, and it looks like it has had a lot of use. Perhaps they travel here periodically for fishing trips, then store up the fish for longer periods."

"I have a report ready to send to Larry and Melvin. Did you send a report on the fishing line tangled in the tree at the South end?"

Six answered, "No, I didn't think of that. I'll send in a report now."

After sending the reports, the time is now eleven-forty-five, and they decided to have their lunch before continuing their search. About half-way through lunch, Melvin called Five's phone,

Five answered, "Hi Melvin, just a minute, I'll turn on the speaker so Six can listen."

Melvin answered, "Good idea. I only have a minute. My main reason for the call is to congratulate you guys. That's a great job on both, the campsite with the hut and the fishing line tangled in the tree. Just for your information, Three and Four found a path with a north-south direction. It looks like it is well-traveled. We are now on the lookout for new neighbors. Gotta go for now. Talk to you later."

Three hours later, three o'clock, and members of the North team are completing their assignment for today. Melvin had been given a flagpole and a yellow banner (to match their ballcaps) stenciled with "NORTH" and all seven members were told to gather at the flag when finished. Melvin had assembled the pole about an hour ago. He had expected it would be a little cumbersome to have it open, but after a few minutes he said it was almost like using a walking stick. Not cumbersome at all.

Finally, looking for a place to stand the flagpole, Melvin realized he needed to find an area that would be visible for – how far? – perhaps a half-mile. We needed to find a spot on top of a hill, and with no trees. Our team members One and Two would be approaching from the west, and Six and Seven from the east. Eventually they found an acceptable spot, on the top of a hill, and with a clear view of the east and west.

Melvin phoned team members One, Two, Six and Seven to tell them we were here, on a hilltop and with the flag flying. One said he and his partner just finished their assigned area and were now heading east.

Melvin said, "Let's hang up for a couple of minutes, then I'll call you back with further directions."

Two minutes later, Melvin dialed One's phone. After they were connected, Melvin tapped the button on his phone titled "where are you" while they were talking. Melvin then said, "I just compared my compass to my phone, and you are 256 degrees from me. So, your course to join us is 076 degrees. The time should be an hour to an hour and a half. Now don't kill yourselves trying to hurry. We still have two or three hours before Larry is here to take us home. We'll see you soon."

Then Melvin called Seven and repeated the process

Done. Now they sit and wait. Melvin wasn't pleased with sitting and waiting. He said to Five and Six, "It has been a grand day for us, and I had a short night. I'm in need of a nap, so I'd like it if you two could post a lookout, then wake me when the first group is about here."

About an hour later, at four PM, the first group appeared to the west. One took a bearing, "They are at 256 degrees, so they have been following their compass."

Two pulled the flagpole out of the ground and waved the flag at them. Another fifteen minutes and we woke Melvin, and then we welcomed One and Two. Melvin said, "Let's have a little celebration. Here's a cookie and a half-cup of cold coffee for each of you."

A few minutes later, Melvin asked, "Can you tell us more about the paths you found?"

Two spoke up, "Well, we found the main path about two hours after we started our search. I was steering 320, walking about a mile and a half from the border, and I walked right into it. Like, you know, like someone placed it there for me. At first it was just a small dirt path, then, maybe an hour later, it was wider, maybe five or six feet, and then we saw some car tracks." Jaw drops by Three, Four, and Five. "And then, we walked on a gravel road, probably fifteen feet wide, and shortly before we were finished for today, we were walking on a paved road. Blacktop. And they did a good job."

Melvin commented, "Well, we were looking for evidence of habitats, and you guys certainly found them. What direction was the road headed?

"The road had been headed about 320 for the first half, and then it made a turn to the left, heading about 290 degrees according to my compass."

Melvin asked if they saw any crossroads.

One answered, "Oh yes. About midway of our trip today, there was a definite path that broke off and headed toward the east. I measured it as 110 degrees. We were thinking that path might head toward the lake that you guys were looking at."

Five broke into the conversation, "I can see Six and Seven. Over there, between the trees, at 105 degrees."

Melvin waved the flag, and then we saw them wave back.

Twenty minutes later we all shook hands, and Melvin handed out his serving of cookies and cold coffee. He asked, “Well, what stories do you have for us?” Six, we heard your report earlier. Do you have anything to add to that?”

“Not really. I could add, though, that the campsite appears to have been used many times.”

“How about you, Seven, do you have any exciting stories to tell?”

“No, and I am sorry to say my trip was without event. No, wait, I found a hat. Early in the trip I saw something moving up ahead. Moving from right to left. At first, I thought it might be an animal. Maybe a fox, or a wolf. I took out my handgun, turned off the safety, and I was ready to protect myself. But then, as I got closer, I saw it was a straw hat. I put the handgun away. And you know what? I was actually embarrassed. I could feel my cheeks flushed. But there was nobody around, so what did it matter. Other than that, I have absolutely nothing to report. Oh, here is the hat.” He opened his backpack and took out the beat-up hat.

Melvin commented, “You should hang onto the hat, and when we are back at your ship, you can give it to Larry. I don’t know if it’s important, but he will let us know.”

A few minutes later, we heard the tell-tale buzz of Larry's buzz-craft. He circled around us once and then landed. Larry was all business. "No time for discussion right now. Everybody, jump in please. This is the five o'clock pickup. I will be back here at six, seven, and eight to pick up the others. Then we'll have a wrap-up meeting at nine, and then I'll buzz Norm and Melvin for home at ten. This will be a long day for all of us."

While the others were finding their seats, Melvin was telling Larry about the blacktop road that goes off to the northwest.

Larry said, "We'll take a quick look in that direction after takeoff."

In the air, leveled off at a thousand feet, they found the road right away. And then it disappeared under a tree-forest. Larry moved us up to two thousand so we could view a larger area. We spent five minutes looking for the "other side" of the trees, but we never found the road. Larry said, "Sorry folks, we must go. We can find a way to check this out tomorrow."

Larry had the throttle close to full forward and then back a little when we reached our cruising speed of four thousand MPH. Thirty minutes later, we landed in front of SC253.

As we stepped out, Larry said Jim Womack, his operations officer, would show us to the same rooms where we had worked the day before.

Jim said, "Hi Melvin, we will go up to the 10th floor. Then you guys can pick up your clean clothing, and I'll show you where the shower rooms are. Take your time, and I'll have your dinner waiting for you when you are ready."

It was a casual dinner. We finished in just a few minutes and loaded the dirty dishes onto the dish cart. Now, back to our business. We arrived back here at five-thirty. The other three teams will be arriving hourly, one team at a time, with the final team expected at eight-thirty. Then we will have our wrap-up meeting at nine-thirty, and we should be done by ten o'clock.

Jim had more instructions for us. "Let's take advantage of this slack period to finish our reports. The tables are hooked up so you can look at the earlier film on the areas you searched. When you write your reports for today, try to reference today's events to those areas already itemized in the films. If you have any questions or need anything, you can find me at one of the tables. Oh yes, we have coffee, several other refreshments, and some snacks, and you can help yourselves anytime."

Melvin is a good manager for our team. Five members from SC253 were recent college graduates with a major in geology. Melvin was very helpful, giving hints on what was important and other suggestions for preparing their reports. At one point, one of them asked, "Gee, Melvin, I would be nowhere without your guidance. You have spent most of your life living inside your space cruiser. How is it you know so much about all this?"

"OK. Short story here. In early college – yes, we had college-level courses - I initially thought I would go for a degree in math or physics, and I included some courses in other subjects, just to test the waters, so to speak. Well, I fell in love with geology, and I found I had a good knack for it. I ended up with a masters' degree in geology, with minors in math and physics. After college, I started teaching math and physics to high school students. After four or five years, they were looking for someone to teach a college course in geology. A few days later, I talked to the recruiter, and then several of the current professors. They asked me a million questions, and then made me take a series of current exams. I passed everything and then taught geology for ten years before we landed on Alpha-1. And now, for the past ten years, I have been the senior geologist for our crew as well as those from SC15, who have been living here for four-hundred years."

"Now, here's a piece of advice. You guys have less than two years of practical experience, and you are probably called newbies; but I don't want you to ever stop learning. Every time you start a new task, you should consider this as a new learning experience. You need to take advantage of these occasions, so you can keep learning. And what about doing the same task day after day? You can use these opportunities to refine your techniques. Think of moving yourself from good, to better, and then from better to great."

The East team arrived at six-thirty. We barely had a chance to say howdy and they disappeared for shower and change of clothing. After they finished their dinner, they paid us a visit and we exchanged stories about interesting events. The only interesting item was the small river that flowed in the east direction in the middle of their quadrant. They found several indications of people fishing from the riverbanks.

The South team was the only team that found nothing worth reporting.

The West team found the blacktop road near their northern border. The road appeared to be heading northwest when it left their assigned area. When Larry picked them up in the buzz-craft, they spent a few minutes looking for the road to reappear on the other side of the trees but were not successful.

Larry called for a brief meeting, with the four teams. "Team West, we'll ask you to postpone your shower. I promise, it won't be more than a few minutes."

Larry continued, now talking to the four search teams, "OK guys, we have almost definite proof that there are people living just off the northwest corner of the areas we searched today. We have photographs of a well-used campsite at the northern part of the lake, which is in the center of our North section. We also have photographs of a blacktop road in the northwest part of the North section, The road left their western border and passed through the northern part of the West section."

Continuing, "Now, do we have proof that people live here. No, the people who did these things might have died off, or maybe they moved to another location. I want to take a small number of us in the morning. The buzz-craft has wheels, so we can land on the road just before the trees start. Then we can drive the buzz-craft at a slow speed and see what we can find. I will bring the side-cart just in case we want to take someone for a ride."

That last comment caused Norm and Melvin to turn heads, as if to say, "And what would that be for?"

Larry replied, "I don't expect it will be needed, but it really simplifies things if the unexpected happens. And it is absolutely no extra burden."

Day 3, Found the Survivors. Larry and Jim showed up in front of Melvin's house at seven-fifteen. Norm and Melvin are standing on the porch and ready to go.

Larry is anxious to get started, so there is very little conversation. Larry lifted us straight up to five hundred feet and quickly accelerated to two hundred mph, then increased to four thousand. Thirty minutes later we are over the blacktop road. Larry set the buzz-craft to head for the northwest, 315 degrees, and a speed of one hundred.

Larry said to his team, "I'll keep us at a thousand feet so we can have a good view in case we find areas with fewer trees. Then when we come to the other side of the forest, we can look for the road again. If all that fails, then we'll put the buzz-craft on the ground and pretend it's a car."

We found several areas where they could see the ground between the trees. Larry took us down to two hundred feet where we had a really good look at the ground. There was nothing to see. Just grass and other undergrowth.

After an hour, we went back to the road. Larry tapped the wheel button and made a gentle landing on the blacktop. A few minutes later, Jim spoke up, "Larry, you're driving pretty slow. Will this go any faster than ten miles an hour?"

Larry chuckled, "You know I get used to the high speed when we're in the air, then when I want to taxi on the ground, it takes a few minutes before I am comfortable with it. The buzz-craft will go much faster than we need. I'll keep us in the 20 to 25 range for now."

"By the way, if you happen to hear gunshots stay down low. The sides of this are bullet-proof. The glass is supposed to be bullet proof too, but I'm not anxious to test it. If we have to return fire, I have rifles and side arms in the back-seat compartment below your feet. I don't expect we will need them, but I really don't know what their reaction might be when they see this strange vehicle in their neighborhood."

A few minutes later, we saw a vehicle coming toward us on the other side of the road. There was a man driving the car, and no passengers. Larry waved, the man returned the wave, and that was that. Larry asked, "Did anyone take a picture?"

No response, other than one low voice, "Sorry. I was looking at the car. It certainly did not look like a Ford or Chevy."

A few minutes later we saw two houses. One on each side of the road. Another half-mile, and we are at the edge of a town. Nice houses. Lawns well kept. This was a big improvement over Our Town. Then we saw sidewalks, and a few cars parked along the edge of the street. Larry parked the buzz-craft, and said to Norm, "I'm going to step out and ask someone where I can find the Mayor's office. Would you and Melvin like to go with me? We can leave Jim here to watch over the buzz-craft."

"Good idea. What if they ask where we're from?"

"Let's give a truthful response. How about, Two thousand miles to the west. We're on a research mission, looking for other people on the planet."

On the sidewalk, just a few steps from the buzz-craft, they stopped a man and Norm said, "Excuse me, we are looking for the mayor's office. Can you help us?"

"Well, I don't know anything about a mayor, but the city manager's office is in the next block on the right. Keep walking and you can't miss it."

We thanked the man and found the office. As we stepped inside, the receptionist said, in a nice voice, "Good morning. Can I help you?"

Larry replied, "We are from out of town, and we were hoping we could talk with your city manager and exchange information."

"I believe you are in luck. He is having a slow day today. Just a moment and I will check." She disappeared through the door." A moment later she opened the door and said, "He will see you now.

Please step in. He wanted me to join for taking notes. Is that OK with you?

"Certainly. And thank you."

The city manager stood up and said, "Please come in and have a seat. My name is David Parker. Please call me Dave. So, tell me what brings you here, and what can I do to help you?"

Larry nodded for Norm to answer this question.

Norm answered, "We live in a town to the west of here. We have always wondered if there were others living nearby, and we decided to start looking around."

Dave asked, "I am not aware of any towns to the west of here. How far away are you?"

"Now hang onto your hat. In round numbers that is two thousand miles."

"Oh, come on. It would take half a lifetime for you to travel that far."

Larry answered this one, "The cruising speed of my ship is four thousand miles per hour. A few days ago, we were taking pictures, thinking we might find evidence of other people. Melvin can give you some details on that."

Melvin talked about the cave, why we bothered to look there and what brought us here.

Larry continued, "The next day we brought four search teams to check the area around the cave. The North team found evidence of some pretty serious fishing around the lake, and then we found the road. Then we came back this morning, anxious to see where the road went, and here we are."

Dave replied, "It sounds like you have a lot of resources available for all this. Please explain that."

Larry continued, "We know your ship landed here in the year 2068. Yours was the third space cruiser, SC16, to launch from planet Earth with a mission on this planet, Alpha-1. The first ship, SC12, was launched in 2065. It had several problems, one making the Alpha-1 destination not changeable, so they were not able to return to Earth for repairs. Another problem was they lost communications, and NSEA headquarters assumed the ship drifted off and everyone on board perished. Two years later, another ship, SC15, was launched, with the same mission as SC12. Everything went as planned, and the ship successfully landed here some eight or nine months later. Communication back then was slow, and it took several weeks for a routine message to go from here to there."

"NSEA was waiting patiently for more messages to confirm that everything was AOK with SC15. The SC15 crew had been sending messages to headquarters saying their ship was not flyable and they asked for a crew to fly in and repair their ship.

Unfortunately, NSEA received none of these, and eventually they figured there must have been some sort of mishap in the landing. They then scheduled a third ship, yours, to fly to Alpha-1. Your mission was search and rescue, and your ship carried about sixty people. Your ship was apparently in communication up to your landing, and then communication was lost. They assumed you crashed and perished."

"Dave, I have another person waiting in my . . . in my car. Can I bring him in, and then we can continue our stories?"

Dave replied, "Certainly. And this is probably a good excuse to take a short break. This is quite a story, and I am just dying to hear more." Dave turned to his assistant, "Alice, would you be kind enough to see that we have coffee for everyone?"

Alice answered, "Yes, and I'll be right back, so don't start without me. You know, I must keep reminding myself, this is real. Wow. What a story."

Larry phoned Jim, "You can bring the buzz-craft to the front of the office and then come inside to join us."

After introductions, the group was anxious to hear more.

Larry continued, "I am the pilot for the fourth ship, SC253, with the primary mission to look for useful resources that the scientists and engineers back home can look at and decide if there might be some resources we would want to start taking back home. Almost an afterthought, I was assigned a secondary mission to see if we can find the remains of your ship. My orders tell me if remains have been found, then the secondary mission becomes the primary."

"Dave, we are anxious to hear your story, but first I want Norm, here, to tell you what happened to the first ship, SC12."

Dave had a surprised look on his face, "I thought you said it drifted off and everyone died."

Larry gave a quick reply, "No that was NSEA's assumption."

Now it was time for Norm to tell his story. He cleared his throat and started, "My name is Norman Link, and I am the captain and pilot of SC12. But don't confuse me with the original captain. He was flying the ship four hundred years before me." Norm gave short versions of the almost unbelievable story. He explained about the two major problems, the first of which prevented us from reaching our planned cruise speed, and the second that would not allow us to change our destination. If we wanted to live, we had to put up with our 400-year schedule. That's twenty generations."

"The ship was equipped with a 100 percent recycle system, and an extraordinary kitchen system for putting together fabricated meals. He explained,

"The rest was up to us. We started with a management system that included a board of directors to put together systems that would work. We had been very serious about carrying out our original mission. After all, we had signed a piece of paper by which we committed ourselves to do our best to carry out our mission, SO HELP ME GOD."

"We started explaining our intentions to everyone, and they, eventually, signed our new commitment papers. We all knew of course we would have to start producing offspring, and then educating these young ones in the same, or similar, academic fields of their parents. Creating an education system was not difficult, because more than half of our personnel were retired or current teachers or college professors. Most of the others were current professionals working in the fields we needed to perform our mission,"

"We spent a large amount of time studying the operations handbooks so we could teach ourselves to fly. In retrospect, it was more difficult than I had expected. The operations manuals give us detailed descriptions of how to do anything we possibly want to do. But it is a large document, and we can't be flipping pages when something unexpected happens. So. one of our first steps was to memorize the manual. The second step was to start practicing everything. That included all the things we know we will have to do. And it also included some other things we might have to do."

"Then we thought about what is lost if we don't make it. For one thing, we lose the ship. For another, we lose the opportunity to meet our commitment and accomplish what we are there for. And how about the 225 lives? It is much more than that. There are twenty generations of these people, and that amounts to 4,500 lives."

"Sorry if I took so much time for that. I would like for our head geologist, Melvin Sherwood to take a turn."

Melvin described our search for lakes with drinkable water, and our unexpected discovery of a family living in a farmhouse near one of the lakes we found. They were descendants of the crew of SC15, which launched two years after SC12, and with the same mission. "They were very hospitable, and we enjoyed telling our stories. This was a very enjoyable event. There was a town just a half-mile away named Our Town. The family's dad called the mayor of the town and arranged for us to have a two-hour visit with the mayor and his staff the following morning. The town and neighboring farmers all lived in an area not more than three miles across. Being here for four hundred years, their ship's crew and passengers of two hundred and fifty people had now grown to a population of five thousand. The people from SC12 ended up joining those in Our Town, albeit all in the same section at the edge of town."

Larry stood up to tell the next story. "Hi, my name is Larry Anderson, and I am the captain of SC253. We just left Earth about six weeks ago. Our primary mission was the same as SC12 and SC15, to look for useful resources. Our second mission was to be on the lookout for any evidence from SC16, which they lost track of, and may have crashed during landing. If we did find evidence, the secondary mission becomes the primary."

"Yesterday afternoon we found just four or five miles north of the cave a blacktop road that sort of disappeared, under the cover of a large forest. This morning, we landed our buzz-craft and drove as far as your office, here. Dave, would you like to tell us your story now?"

Dave stood up and said, "Yes, I would like to do that, but first I'd like to see what a buzz-craft looks like. Can we take a look?"

"Of course. It is parked out front."

Larry gave them a good description. It carries four, and there is an optional side car to carry up to six more; top speed over seven thousand mph; excellent cameras with high quality pictures; programmable. "These were first introduced about three years ago. They had a problem right away with idiots running into each other. The FAA grounded all of them, and then started creating a bunch of flight rules. Then they required everyone to take flying lessons, and then pass a written test before they could apply for a flight license. About half of those crafts that people bought went up for sale. They didn't all sell, so that brought the price down and that is when I bought this one."

Back inside, Dave now had the floor. "I certainly hope you don't expect me to have all the details of our mission. The best I can do is a brief summary for the flight of SC15."

"The leaders at NSEA were concerned about no response from SC15. NSEA had received a message, "preparing to land" or something like that. They waited for two months, and then said they had to assume it was a crash landing. They then decided to launch SC16 with the mission of search and rescue. Our Captain was new. He had received a flight license but did not have much experience. All appeared to be OK up to our landing on Alpha-1. The new captain was so nervous he fainted. One of the other members took the captain's seat and said he could make the landing. Unfortunately, he had the left and right buttons reversed and ended up landing the ship on a hillside. The ship stood there for an hour or two and then tumbled over and rolled the rest of the way down the hill. Fortunately, they had enough time to abandon ship, and no one was hurt."

“The crew was afraid of a fire, and they removed everything they could. The technicians said the engines were damaged beyond repair. The communication equipment appeared to be ok, and they started sending messages. We quickly built a series of lean-to huts to live in, and then settled down to wait for rescue. After waiting a year, we decided to start planning our lives on Alpha-1. We had a crew of sixty people, and only fifteen of those were females. Initially, we thought we should not have babies, because that might be a problem when the rescue ship arrives. But after waiting for a year, almost all of our ladies agreed to have as many babies as they could. They did well, and over a fifteen-year period, our first generation produced forty-five babies.”

“We were fortunate to have a ship full of people who were always thinking about improvements to what we had. We started with sixty bright people and now we have grown to eight hundred. In the beginning we were satisfied to live in large buildings that would each handle up to a dozen people. But more recently, most of our newer homes are pretty much for single families.”

“Our residents look forward to improving our lifestyle and are always offering suggestions. We built a small factory at the twenty-five-year point. The first products were tools, like hammers, saws, and other small tools. Then we moved to lawn mowers and bicycles. Then we had motorbikes, and in another five years we started producing simple styles of cars. Today, our cars are still simple, but they are mechanical wonders, and they're quite attractive. And yes, we have thought about airplanes. We decided to put that off for a few more years."

Dave sat down and took a sip from his cup of cold coffee. Alice immediately refreshed his cup. Dave said, "I forgot to mention we have an education system. Education was almost forgotten in the early years. Then, as our babies were becoming five or six years old, it was a definite reminder. We expanded our education system as our children reached ages that were beyond the limits of our existing capabilities. Our current grade school and high school systems are working just fine. College is a problem. Recent high school graduations usually have about nine or ten each year, and maybe half of them are interested in some college courses. Interesting, the ones most interested in college are usually our high school and grade-school teachers. These are all bright people, and several times we have seen current teachers taking college-level courses. The teachers pick a course to teach, study-up on the requirements, and then go to work teaching other teachers."

Norm looked directly at Dave, and said, "Wow. What a story. You - I mean all of you, starting with your ancestors - have obviously done a remarkable job of adapting to this horrible nightmare." Now, Norm looked at Larry, smiled and said, "Larry, I believe the next steps now fall on your plate. Over to you, sir."

Larry turned to Dave, "Yes, according to my orders, this is now my primary assignment. I have already sent two reports to NSEA that should be of interest to you. The first report was two days ago and had to do with the painting we found on the cave wall, and empty spray cans on the floor. The second report described yesterday's recorded evidence of frequent fishing at the lake north of the cave, the lean-to hut, car tracks, and the blacktop road. That report also mentions that we plan to follow the road this morning, Travel time for reports is shorter than it used to be, so they should receive my first report today. I do not know what they will want to do, but I recall seeing something about wanting to return everyone to the headquarters location."

Dave had a worried look on his face. He said, "Now I am a little worried. No, maybe a lot worried. Are they going to make us pack up and return to Earth? SC16 was the third ship stuck on Alpha-1. What was their reaction to the other two?"

Larry took his time to answer Dave's question. "Well, my orders tell me specifically what to do on this mission, and that includes reporting the offspring of the SC16 passengers are alive and well. My orders did not mention the other ships. So, what shall we do? I can't imagine that NSEA would want to order you to return to Earth. For one thing, they probably do not have the law behind them. I wonder if laws about these things even exist."

Norm was looking agitated. He cleared his voice, "Regarding the SC12 crew, I leafed through the minutes of annual meetings, and it is clear they all felt duty-bound to carry out the mission assigned to our forebears. They signed agreements before the flight that they will do their best to carry out the assigned mission, SO HELP ME GOD. This agreement was also reviewed at each of the annual meetings, and passed on, generation after generation. During our last year of the flight, we had frequent meetings to discuss our specific duties, and we started preparing action plans. Later, perhaps a year or so after our landing, and repairs were performed on our operations computer, we started having troubled thoughts about a return flight."

Norm's face was turning red, he stood up, and his words carried a lot of energy, "And now, ten years after landing, I believe nearly all of us would prefer to stay here rather than risk a return flight. After all, how do we know the repaired computer is fully repaired? We would have to be airborne to give this a full test. And who wants to take a ride in a space cruiser that is four hundred and ten years old and has such a bad history? And on top of that the communications equipment was still not working. All considered, I, for one, have already made up my mind, count me out ! I will do what I can to stay right here."

Norm sat down, looking exhausted.

Larry was calm, cool, and collected. He waited for Norm to calm down, then he offered this, "Norm, I would expect that you and your crew will not be required to return to Planet Earth. I don't know these things for sure, but I would expect NSEA will be very lenient with you and the other ships' crews. Perhaps they will offer you options. One option might be an all-expenses-paid trip back to Earth. Another option might be for NSEA to provide a home improvement deal, where they provide to your home, or to your neighborhood, your city, a general fix-up program. I heard somewhere that a single round-trip to Alpha-1 might cost around ten million dollars. I also heard somewhere that NSEA is thinking about establishing another NSEA complex on Alpha-1, that could serve as a jump-off point to other destinations in the universe."

Larry's comments really got everyone's attention. Dave was excited and he spoke up first, "You know, this is probably the most exciting two hours I have ever spent, I am only forty years old, so these are things that can still happen in my lifetime. Just think, I could actually see historic events happen. And maybe I can be part of these events. As you may have figured out, we have been living a pretty dull life."

Norm then leaned forward and said to Dave, “You heard Melvin mention settlements. He is two thousand miles west of here. I was wondering if you, as the manager of this city, would like to meet with the managers of Our Town and West Village? We can probably talk Larry into providing the transportation. It’s only a thirty-minute flight in his buzz-craft.”

Dave was nodding his head, and then he said, “Yes. Sounds good to me.”

Larry jumped in, “First, I need to make a call, to see if those folks can be available for a mid-afternoon get-together.” Larry’s “special” phones provided a direct connection only to the staff on Larry’s ship, SC253. Once connected there, the other person forwarded the call to Maynard’s office.”

Two minutes later, Larry heard Maynard’s voice from the other end, “Hi Larry, how’s your search going?”

“Great. We just spent two hours with Dave Parker, the City Manager here, and I am wondering if you would be available to visit with him this afternoon, say around two PM. Nothing formal for today. Just get acquainted, and then we can prepare for something later on.”

Maynard said, “How about I bring along our two mayors?”

"That will be great. Say, we're going to be hungry. We can be there about one. How about us stopping at the restaurant across the street from your building for a quick lunch?"

"No, I have a better idea. We will be meeting in my large conference room. I will have lunch waiting for you. How many are you?"

"Maynard, that's just great. And there will be eight of us."

"All that works just fine. See you at one o'clock."

Larry turned to Dave, "We will need to leave in about fifteen minutes. Can you quickly grab three others for the trip?"

Dave nodded in the affirmative and picked up his phone. Five minutes later, he said, "My team will be here any minute now."

Following brief introductions, all eight are strapped in, and we are underway at twelve-fifteen.

Heading east on the blacktop road, Larry kept increasing the speed. At thirty miles per hour the vehicle was bouncing a bit too much. Larry slowed, "Sorry about that," then slowed to a more comfortable twenty. A minute or two later, we are at the edge of the forest and Larry said, politely, "Hang on, now we can go a little faster."

Larry hit the hover button; the ship started upward, and the bouncing was gone. Then he started moving the throttle forward to increase speed. A minute or two later, he said "We're doing four hundred, no, five hundred now, and going up to four thousand. I have made this trip a dozen or so times now, and it takes about thirty minutes. Is everyone OK? It's awfully quiet in here."

They remained quiet. They are not used to this rapid acceleration. Larry and his crew were grinning. The other four were still very quiet.

Larry said, "I'm slowing us down to three hundred and we'll take a quick tour of the neighborhood. Norm, why don't you be the travel guide for our guests."

Norm perked up, and said, "Certainly. We are a few miles east of Our Town. Coming up on your left is the smelter and after that is our larger manufacturing building. When my ship arrived, ten years ago, they had just moved from bicycles to motorbikes. Now they make cars and helicopters; and maybe manufacturing airplanes in a few months. Just ahead is Our Town, the home of most of the people from SC15, now some twenty-five-hundred people. They also have another hundred and fifty who chose to live on their farms. I believe all of them are within ten miles of the town. You will see a group of newer homes at the western edge of town where my crew of two-fifty live. We call this Western Village. Now, left turn here, and you will see Larry's ship, SC253, standing in a cleared area. Then we swing back to town. Let's all make sure we are buckled in as we approach our landing spot on top of this building."

To our surprise, the governor's group was standing on the rooftop, at a safe distance from the landing pad. They were all waving like crazy, so it was a nice welcome. As soon as the "buzz" stopped, they all came up closer so they could get a close look at Larry's buzz-craft, and our new guests from SC16. Maynard Jones, Governor of the combined areas, ushered all of us down the stairway and to his very comfortable conference room,

Key personnel attending included Maynard and his personal staff, Tommy Smith, Mayor of Our Town, with several of his directors, and Tom Kennedy, Mayor of West Village, with several from his staff.

Maynard is the self-appointed master of ceremonies, of course, and he does a pretty good job. Lunch is the first thing on the agenda, and we have three choices for the main course: roast beef, venison, or fish. While the kitchen staff is starting to serve the loaded plates, Maynard started his version of SC15's landing four-hundred and eight years ago, and their wasted first year waiting for someone to rescue them.

Maynard had given this spiel several times, "The hardest part was accepting the fact that we are stuck here, and we had better get started on making the best of it. A few of our crew volunteered to take up farming. The internet system that was provided in our on-board computer network was very helpful for these things. The internet also helped us learn how to search for iron ore, then the smelting process, and then making useful products. We also learned how to build houses, create electricity, provide drinking water to our homes, and many other things. Oh yes, our computer experts didn't need much help. They were already experts in their fields, so we had a "wired" computer network in a couple of months, and fully wireless by the end of our second year, and a phone system about the same time. In addition, we eventually started a small factory for manufacturing. Their initial products were small tools, then we had lawn mowers, bicycles, and they started receiving more requests than they could handle. Business was accelerating. Bicycles led to motorbikes, then motorized tools, cars, trucks, and helicopters. We have been working on airplane designs and we should soon have a real airplane."

At that point, Maynard announced, "The lunch period will be closed in five minutes. Then we need to stay out of the way for a few minutes so the kitchen personnel can clean up from our lunch, then we can re-join."

Fifteen minutes later. everyone was again seated at the table. Maynard returned to the podium, "Ladies and gentlemen, for four hundred years there has been another ship on Alpha-1. Just two thousand miles away. Here to tell their story is David Parker, City Manager for the eight hundred individuals who are the twentieth generation from SC16. Dave, the podium is yours."

"Thank you, Maynard. You know, just this morning, we met the group looking for us. This was an absolute shock for us to learn of your existence. You were here for those four hundred years, and we never knew it. So, I have now heard the stories regarding SC12 and SC15. Those are exciting stories. SC12 and SC15 both had the mission of looking for an alternate source for some of the basics that are now dwindling on Planet Earth. SC16 was different. Our ship and some of the crew had been preparing for a similar mission, but to a different planet. When NSEA finally understood that SC15 was missing, they decided to send our ship on a search and rescue mission. SC16 launched in 2068 and all seemed to go as planned until the landing. "

"The captain of SC16 had only recently completed his flight training. During the landing phase on Alpha-1, the new captain was so nervous he fainted. His back-up took over for the landing, Unfortunately, he got the left and right controls reversed and landed the ship on a hillside. The ship stood there for an hour and then fell over and rolled the rest of the way down the hill. Fortunately, everyone had enough time to abandon ship, and no one was hurt."

"They were afraid of a fire so they removed everything they could. The engines were damaged beyond repair, but the communication equipment appeared to be OK, and they started sending messages to NSEA. After a few weeks, they eventually realized the connection with NSEA had been lost."

"They quickly put up a series of small buildings to live in, and then settled down to wait for rescue. After waiting a year, they decided to get started with planning their lives on Alpha-1. They only had a crew of sixty people, and only twenty of those were females. Our families did well, and within the next fifteen years, our first generation had produced forty-five babies."

Dave took a relaxed posture, and said, "I do believe the remainder of my story would be about the same as the others. Maynard, the podium is yours."

Maynard thanked Dave for his presentation, and then, "The next part of our agenda should be very interesting. Larry Anderson, standing here, is the Pilot and Captain of SC253, the ship that landed five miles away from here, just a few days ago.

"Hi, I'm Larry Anderson. As Maynard said, I am also the pilot of the buzz-craft you have seen flying in and out of your town a few times. That is a handy vehicle for lots of things, and it fits just fine in one of the cruise ship's storage areas. With me is

Commander Jim Womack, my Operations Officer and number one assistant. Our total crew is one hundred and twenty people, but you probably won't see any of the others. Our mission started out to be just like the first two ships; that is, to survey the available natural resources that might be available to offset the shortages that are developing back home. Then, just two days before leaving Planet Earth, the boss asked me to make a quick survey of Alpha-1 to see if there are any signs of SC16, which had a search and rescue mission four-hundred-plus years ago, looking for SC15. "

"OK, here is a quick review, and I will point out the assumptions made by the senior staff at NSEA. They launched SC12 in 2065. SC12 had problems. Then they lost communication, and NSEA assumed the cruiser drifted off in space and everyone died. Now, four-hundred-plus years later, we know that those on SC12 somehow stayed healthy and survived the four-hundred-year trip; and they landed here in 2465."

"Having lost SC12, NSEA launched a replacement, SC15, two years later, in 2067. Everything had been going fine with SC15, and NSEA received a message saying they were in orbit around Alpha-1 and preparing to land. That was the last message received from SC15, so eventually NSEA assumed their ship crashed on landing, and all on board must have perished because there was no message."

"NSEA was duty-bound to send another ship for search and rescue, so they hurriedly arranged a change of plans for SC16, which had been preparing for a different mission on a different planet. Like events for its predecessor, the flight went as planned. But unlike their predecessor, they had a bad landing, and the ship was a total loss. Everyone vacated the ship without any serious injuries. Communications had been lost. The crew built some basic living spaces and prepared to wait for search and rescue."

"NSEA went into panic mode. Two ships were lost, and only a year apart. How or why did this happen? Was there some strange magnetic force that interfered with the landing, and then maybe interfered with their communications? Was the planet inhabited with their own population that did not want us to land? Were there aliens from another planet that did not want to share anything with us? The last thing they wanted to do was launch another ship that would also be lost. They decided to sit on it. Perhaps they might somehow learn something later. They sat on it, and then they forgot about it."

"Four hundred years later, now three days ago, I had the pleasure of meeting Norm, Maynard, and Melvin, and we spent a few hours reviewing photos and picking some areas to look at. The next day, Melvin, the lead geologist, said he would like to look at the inside of a cave that happened to be in the second or third area we looked at. Melvin said he would be looking for wall-paintings, which could give us evidence of prehistoric people. Well, he found some wall-paintings, and he was very excited. He took some pictures and was heard to say, '*We will be famous*!' But in his pictures, we also saw used cans of spray paint on the floor. Melvin was devastated. He hung his head and said, "I'm sorry, we just wasted our time." But this was not time wasted. This was our first evidence of other humans on Alpha-1. This was a good day for us. And it also changed my mission to search and rescue."

Larry continued, "That same day we identified the areas for the next day's search. The four pie-shaped areas were centered at the cave and extended out for five miles. The north area had their first success about two miles from the cave, and at the southern edge of a mile-long lake. This was a fishing lure caught in the overhead tree branches. More evidence. Then, closer to the north end of the lake was a semi-permanent campsite, and a mile further we found a gravel road and car-tracks. Now we really got excited. One of our team followed the road further. Not very far, it became a blacktop road, and it was headed in the northwest direction. About this time, I showed up with the buzz-craft. It was late in the afternoon, about five o'clock, and we and had to head back to our work room on SC253."

"The next morning - today - four of us in the buzz-craft headed for the blacktop road. As we approached the tree area, I tapped the wheels-down button and landed on the road. We drove on the road for maybe fifteen minutes, and we were in the edge of a small town. We decided to go for a walk and see what we could see. We left Jim to watch over the buzz-craft. A few minutes later we met a man on the sidewalk, and we asked if he could tell us where we could find the mayor's office. He was helpful and just a block later, we met Dave Parker, the City Manager. We had a good conversation there. Then we made a few phone calls, and here we are, in Governor Maynard Jones' office."

"During our visit with David, there was a question that needs to be addressed: Will NSEA try to force us to return to Planet Earth? There is a clause in all of these agreements for personnel who will be landing on another planet. The clause reads something like this: If an employee is lost or stranded on another planet, NSEA will be duty-bound to send another space cruiser to locate the individual or individuals and return them to Planet Earth ASAP."

"I understand some of you would rather stay here than return to Planet Earth. I will forward this information to NSEA. I was thinking about including a suggestion that NSEA could offer options to those stranded on Alpha-1. Perhaps we can have an open discussion on the options, today, this afternoon. Then after this meeting I will prepare my third report to NSEA, and that will include these suggested options."

The next five minutes involved some of their people jumping up and sounding off. Larry asked for a hand-count. Everyone was cooperative for this, and the count was twelve for staying on Alpha-1, and seven returning to Earth. Several of those present chose not to vote.

Norm remembered what Larry said in David's office this morning, and he raised his hand. Larry saw whose hand this was, and probably was thinking, please don't get me in trouble. He said, "Yes. Norm, would you like to add something?"

Norm stood up and said, "I am certain that sending a number of spaceships to Alpha-1 for returning all these people back to Planet Earth would cost NSEA a lot of money. So, with two-thirds of us electing to remain here, we are saving NSEA a lot of money. Maybe we could have a compromise solution and split the difference. I don't think we should be asking for money for our personal benefit, but rather for improvements to our living conditions, such as better roads, educational buildings, a library, and so on. Do you believe NSEA might be agreeable to such an arrangement?"

They had a good discussion, but nothing new. Just emphasizing their desire to remain on alpha-1, and their question about some big business activities opening offices here and encouraging more citizens to move to Alpha-1.

Larry gave a sigh of relief, and replied, “Norm, I believe that is a great idea; and yes, NSEA just might be agreeable to this.”

David put his hand up. Larry said, “Yes David, what do you have for us?”

David stood and said, “I was wondering what we can do to attract big businesses to move to Alpha-1. And also, a related question; what can we do to attract more people to move here?”

Larry replied, “Personally, I can’t think of anything to add to this, but I will certainly include your questions in my report.” Then he turned to Maynard, “Maynard, I believe we have accomplished everything on my list. I have nothing else for today.”

Maynard jumped up, anxious to step back into the limelight. “This has been a very interesting day for me – and for many of you, I am sure. Thank you, David, for telling us about your trip to Alpha-1. I am so sorry your landing was not successful. And I am sorry we never knew you were here. One of these days we will have airplanes, and then we can fly over and visit for a spell. And thank you, Larry, for telling us about the search for the twentieth-generation survivors of SC16. Thank you also for handling our discussion session to support the many people who are happy with staying right here. And thank you, everyone, for participating in our meeting today. That is all for today."

We all stood, and I thought for a moment we were headed for the roof so we could return to the buzz-craft. But, no, it seemed like everyone had to shake everyone else's hand, and share some personal feelings on this and that. After a good half-hour, Larry called out in a loud voice, "OK folks, time's up. Lift-off will be in five minutes." Larry headed for the stairs, and we followed right behind him. A minute or two later, the rest of the group followed, then stood at a safe distance and waved their goodbys.

Larry pointed his buzz-craft east and leveled off at a thousand feet, "This altitude gives us a horizon at about forty miles as well as a good view of the passing scenery below. Dave, if you have any additional comments or suggestions this is a good time to talk. When I get back to my office on SC253, I will be preparing and sending my next report to NSEA. Before we drop you, can you give us a ten-minute tour of your lay-out?"

Dave replied, "Certainly, and we won't need all of the ten minutes. You have been on the main road already. We continue that road, and we see the education buildings, a good size swimming pool, a football field, and several churches. To the left is a large residential area. To the right is one more street with houses, and then another street for factories. And that's about it."

Before long, they were on the road and Larry slowed the buzz-craft down to a comfortable speed.

Dave started his tour, "As you know, this is the main road. On our left are three more streets, parallel to this one, and most of our people live there. Now, straight ahead is our downtown area with shops, two restaurants, and some business offices. Up ahead on the left, the nice-looking brick building is our high school. Beyond that, the rather large older-looking building has plenty of room for junior high and grade school. On the top floor we have several rooms reserved for college-level courses. Across the street from there are two churches and a small library. Several of our people are prolific writers and they say their goal is to fill up the library. You can turn right at the next street. One street over from the main road has more homes. Turn right again at the next street. You will see a few houses on the right. They will be removed in another year or two. Now, on the left, we have several factories. This nice-looking building is our latest factory. That's where we are currently building helicopters and will also be building our first airplane in a few more months. This next building has been here for two hundred years, and they build just about everything, including automobiles. The last building is mainly for plastic. That used to be our smelting plant, but it had some terrible odors, and sometimes a lot of smoke, and it could be noisy, so we built a new smelter about a half-mile to the west. Turn right again here, then stop at the main road."

We chatted for a few more minutes, and then we are on our way. Thirty minutes later, Larry dropped Norm and Melvin off at Western Village. It was nearly eight o'clock and everyone was tired. It has been a long day.

As soon as Norm's wife heard the overhead buzzing, she left the house and started walking toward Melvin's place. Norm and Melvin received hugs and kisses as soon as they stepped out of the buzz-craft. Larry was anxious to move on. He hollered, "I'll check in with you tomorrow."

Norm hollered back, "No. Wait a minute,"

"OK. What's up", begged Larry, "I have a few hours ahead of me."

"Melvin answered, "The four of us are going to sit and chat for a few minutes. Why don't you join us. We'll limit this to fifteen minutes."

Larry and Jim put heads together briefly, then Larry turned off the key and answered, "OK, fifteen minutes and no more than that."

Melvin invited the group of six to come inside. Within a minute they were all holding a glass of wine. Melvin held his glass up and said, "I have a toast. This is to the people who landed these three ships on Alpha-1, and the twenty generations of their children and grandchildren. I cannot even imagine having to live through anything like what they have been doing. This toast is to those of the NSEA space cruisers SC12, SC15, and SC16, obviously the toughest people I have ever met." We all held up our glasses. "Hear, hear," said all six in unison. Then five of them downed a swallow from their glasses."

All but Larry, who said, "Sorry, I can't touch the stuff. I'm driving." That brought a chuckle from the rest of the group. Then he said, "I would like to give a great big thank you to Norm and Melvin for your excellent help for the search. Yes, I could have done it without your help, but the search might have been a couple of weeks instead of a couple of days. Thank you. If I were not driving, I would have made that a toast." Then, before anyone could say anything, "And now, Jim and I must go. Like I was trying to say fifteen minutes ago, I'll check in with you tomorrow."

They all waved their goodbys and Larry and Jim were out of sight in less than a minute. This was a long day for Larry. He stayed up late to start preparing his notes so Jim could help with the formal writing in the morning.

Day 4. Larry's first task was to bring Jim up to date so he could fully understand the messages he wanted to present to NSEA in the report. Larry had several pages with scribble notes. Jim will smooth these out and add them to the first draft. The next step includes preparing a reasonable format for this important report to NSEA. Larry and Jim agreed quickly on the details to be included in the report. The previous report, sent just yesterday, described the evidence we found of people living in the area, with the blacktop road and tire tracks. That report also mentioned our plan to continue the search this morning by following the road. Larry prepared a list of the topics he wanted to cover, "These are the topics for this report. We found a small town, with a population of eight hundred. The layout of the town is very professional. They have several factories, one of which manufactures automobiles, and is contemplating adding a helicopter and an aircraft later this year. We had a three-hour visit with the city manager and three others. This was followed by a two-hour visit to introduce these same individuals to the leaders of Our Town and West Village."

Jim had a few questions. They were good questions, and Larry provided good answers. Larry recognized that each of these question-and-answer routines was adding more strength to their report.

Some new suggestions followed. If people are stranded on a distant planet, NSEA is currently bound to send another space cruiser with a search and rescue mission. Our new thinking was in favor of flexibility on this subject, and options. If these people have been there for an extended period, they might prefer to remain there. Perhaps NSEA could offer an option for them to remain here and provide compensation of perhaps one-half of the cost saved by NSEA to go toward improvements to the facilities where they are presently living.

Larry and Jim work well together. Larry is the pilot and the best engineer in the business, but he is not a good writer. Jim is only an average engineer, but he can work magic with words. As they gathered their thoughts to work on the next piece of the report, Larry was the idea-man, discussing what goes in this section, and then discussing the many pieces, and how the pieces relate to each other. Jim asks questions so he can fully understand all of what Larry wants. When finished, they would part company for an hour or so. In the next hour, Jim prepared his first draft of the text, while Larry was going through his earlier notes for the next section.

Larry came up with another suggestion. "Jim, see if you can add a section to cover this. We can see that life on Alpha-1 would be greatly improved if some large businesses could establish new facilities on Alpha-1 for manufacturing and selling their products. Also, we heard that NSEA is considering a new facility at a distant planet, and we believe Alpha-1 would be ideal for this.

They took a break for dinner. Then they tried to get back to work, but this wasn't working. They were exhausted and agreed to a nap.

Day 5. Morning came, Larry and Jim had a quick breakfast and went to work. A few hours later, they were finally done. They read through the final version several times, Larry finally nodded his head and Jim tapped the send button. Now they could relax.

Chapter 8 – Searching for Resources

Twenty minutes later, they were back inside SC253. Larry led them to their now-familiar work room. “Let’s talk for an hour or so, then we can break for lunch. I’m sure you remember the original plans we prepared about a week ago. That’s what we were doing the day Melvin wanted us to land so he could check out the cave.

“What we will be doing tomorrow is essentially a restart to where we ended up that morning. About half of our effort is looking for iron,

coal, oil, and gas. This is an estimate based more on the terrain than anything else. Your scoring is easy. Three choices: no, possible, and probable. Just check the box, and make sure you take pictures of the terrain. The experts back at headquarters will make the decisions on these, then we will come back with the heavy equipment to do some digging. The other half is looking for minerals. Each team will have a guidebook, with pictures of rocks and descriptions of how to identify various minerals. This is a good book, so be sure you use it.”

“We will have four teams. Each team will have four people. The four of us will be team leaders. I will pick up the two of you at seven-thirty in the morning. Norm and Melvin will meet their teams on the 10th level of SC253 at eight. We will hand out the search areas we agreed to earlier. Then you can pick up your work outfit, phones, and lunch bags. Lift off should be about nine, and I will drop you at your first areas about nine-thirty. When finished, you can move on to your second area. If you need a lift, call me and I'll buzz over to take care of that. You should take a lunch break around one, and then you can move on to your third and fourth areas. Remember, call me if you need a lift. I will plan to pick you up at five and return to our ship."

"We should be back here at five-thirty. I will have to leave right away so I can pick up the other two teams, and then I can buzz you home about six-thirty or seven. That hour should be enough for showers, change clothes, and jot down your notes."

"Teams three and four will have Jim and me as team leaders. Our schedule will be like yours, except adjusted so events start one hour later."

"The next day will be spent right here. I would like to have the four teams together for an hour or two. We'll run down the list to give everyone a chance to tell us about what they accomplished. Then we can break up into four teams in separate rooms preparing a rough draft of our reports. Then I would like to have an hour with just the four of us. We can then assemble the final report. If the four drafts are good, then that last effort might be easy. When finished, I'll send the report and then we'll talk about a planning meeting for the next trip out."

"Here comes lunch. Let's take a break for a while." Half-way through lunch, we started stepping through some parts of tomorrow's schedule. Two hours later, and we're done. Larry buzzed Norm and Melvin home.

Day 6. The next morning, seven-thirty sharp, Larry picked up Norm and Melvin, then we returned to the ship. Norm and Melvin changed into their work clothes, then handed out the area charts and the special phones to their team members. Larry took care of fastening the sidecar to the buzz-craft. He also placed in the rear trunk a set of tools for each team: two shovels, axe, heavy-duty sledgehammer, hatchet, and hammer. He also tossed in a half-dozen heavy-duty bags for samples. Larry had prepared this list several days ago, and the team members thought the list was about right. And then yesterday he thought about how heavy the loads will be and can they manage carrying these tools for the distances they will have to walk. Larry asked the techies in the shop about carts or wagons for carrying the tools. There were none in the ship. But these guys are very handy, and they said they could easily make carts that will work. One of the men drew a quick sketch, and Larry said "Perfect! We'll take four of these."

Larry had parked the buzz-craft in front of the ship and was about finished loading everything when the first two teams appeared. Larry assigned himself and Jim to the front seats, the leaders of teams 1 and 2 to the back seats, and the six other team members to the side-cart.

"Fasten your seatbelts," said Larry. Jim checked over the crew, then gave Larry the thumbs up signal.

"Here we go," said Larry, as he maneuvered the buzz-craft in a rapid climb and quickly to four thousand miles per hour.

Most of the recent college graduates had majors in geology and two of them were mesmerized, watching the surface of Alpha-1 swiftly passing by below. "Did you notice the difference in ocean coverage," said one of the recent grads, "Earth is seventy percent ocean, and this is much less than that. Not even fifty percent. That means those important natural resources should be more plentiful here."

Another grad commented, "All right, so the resources are more plentiful. That doesn't mean they are more available. We are four light years away. That's a long trip to haul a ten-ton load of needed minerals."

"Yeah, but if that's gold, that would certainly lower the price on the open market."

Larry interrupted the discussion, "Hey guys, we're about there, so it's time to fasten those seat belts." Now we are going down. Larry had keyed in the coordinates earlier, and he was able to land the buzz-craft right on the target. The landing point was directly over the cave where we found evidence of other humans a week previous. The area for Norm's first Team started just outside the area of the cave and in the northwest direction. Melvin's Team 2 started at the cave and then continued to the northeast.

The basic training of geologists included searching for metals, minerals, oil, and natural gas. Most experienced geologists eventually picked one or two of these as their specialty, but these young geologists, for the most part, were eager to try out as many areas as possible, so they could eventually decide which areas to pursue. Areas that show signs of upheaval could be the result of volcanic eruption, earthquake, or erosion from streams and rivers.

Larry made a nice landing, and everyone appeared to be happy to have their feet on the ground again. The two team leaders, Norm and Melvin, gathered their teams, loaded the tools in their new carts, and off they went.

Norm and his team members set their sights on the long hillside that accompanied the stream for at least several miles. Their guess was that a serious earthquake caused the terrain on one side of the stream to be fifty feet higher than the other side. But there was no convenient way for them to walk on the higher side, which was fifty feet above the stream, which seemed to have a fairly consistent depth, ranging from three to five feet.

Their first attempt was to walk the low side, looking for rocks in the shallow water. This resulted in a good collection, but after ten minutes the others were just repeats. The second attempt was to wade in the water on the deep side. They tried scraping the bottom with a bucket, and had some luck, but this was the same assortment of rocks. They thought about moving to the high side and hanging over the side on a rope, but they could not find a convenient place where they could climb up the steep slope.

Norm did his best to avoid making decisions since all three of his other team members had just been awarded their degrees in geology. So, he made suggestions, “Hey guys, I seem to remember from the photos, the landscape changes from this to scattered hills, and maybe some smaller mountains. Let’s open up our instructions and take a look. Then you guys should make the decisions. You are the experts. My primary job is to make sure we’re in the right place at the right time.”

From one of the team members, “Well, you are right, the hilled area starts ahead of us, in less than a mile. We should be there in about forty-five minutes.”

Norm said, “Sounds good. Here’s another question. How far is it from here to our second assigned area?”

“Well, here’s a surprise for you. We are already there. It is ten-thirty now and we’re supposed to take our lunch break at one o’clock. It appears we are ahead of our schedule.”

The appearance of the second area is different. The landscape on our left side was a slow rise, perhaps ten feet in the next hundred yards. Just a few minutes later, we heard the sound of running water. “I hear rapids up ahead.” Someone said.

The water was really rushing; and there were rocks of every color and type we could imagine – a much larger variety than we looked at earlier. There are many boulders, the large ones up to five feet in diameter. “Hey Norm,” someone hollered, “This looks like a great area for gathering samples. OK to stop here for a while?”

Norm answered, “Certainly. But don’t you guys get carried away with taking so many rocks for samples. We already have a hundred pounds of rocks in the cart.”

A few minutes, Norm called the team together, “So, how is it going?”

One of the team members answered, “We took your last comment seriously. Our task is to identify the types of rocks we find. We can do this by noting the reference numbers on our worksheets, so we don't need all these sample rocks. Except it might be a good idea to have one of each."

We eventually developed a procedure. Yes, we should have done this earlier. We called the first one Plan A. This will be Plan B. We walk forward in a particular area for, say, fifteen minutes. Then we stop, check the types of rocks here and make our entries on our work sheets. Then we repeat the plan. After an hour of this, we noticed things were changing. The slope became more gradual; and then it was flat, and in front of us is a lake. This appeared to be the remains of an ancient volcanic eruption. Just a few minutes earlier we had seen an area we thought was solid black rock. Now we agreed that was more than likely solidified lava. Beyond this hill we could see several more hills that were larger than this one. Pretty soon we might be climbing mountains.

It is time for our lunch break – although it is apparent that our need is more for rest than food. We only traveled three miles, but some of that was wading through knee-deep to waist-deep water. Some of that was also up hill, The three young guys had been taking turns pulling the cart. Norm just turned 69, but he looks to be in pretty good shape except he's fifty pounds overweight. The cart, with tools, was nearly a hundred pounds. We developed Plan B to save only one of each type of rock we find. That decision reduced the total cart weight from more than two hundred to about one hundred.

We identified a number of new rocks around the lake. The young geologists informed us that a typical volcanic eruption can bring many new varieties from as deep as a mile, and we could expect to see many new specimens throughout the day.

At the far side of the lake, we found another stream bringing water from the two hills that lay ahead of us. The first section of terrain after the lake had a small rise to it. Perhaps three or four feet every hundred yards. After a mile or so of this activity, we came upon a small waterfall. There was a twelve-foot fall and a small lake at the bottom, and then a stream to take the water down the hill from where we had just come. We made note of a few more samples around the small lake. Then we started looking for a way to climb up and around the twelve-foot waterfall.

Carrying our cart up this rise was more difficult than we had expected. It took us nearly an hour to be at the next level and alongside the stream. Surprisingly, the stream here seemed relatively calm. We were now far enough from the waterfall that we could no longer hear the rush of the water. The heavy tree growth cut off a lot of sunlight, providing a tranquility to the area we had not expected. It was time for a fifteen-minute rest stop.

Norm spent a few minutes studying our charts. We had earlier agreed on a pickup spot that was open enough for Larry to safely land the buzz-craft, and we are supposed to be there at five PM. The spot is about half-way up the second large hill, but at our rate we might make it there about six-thirty. Norm came up with a new pickup spot that was at the top of this hill. There was a huge crater at the top, and most of the rim appeared to be free of trees, so that would make for an easy landing. Norm shared his thoughts with the three teammates. They kicked this around for a few minutes and eventually they agreed. Norm called Larry to arrange for this change. No problem.

Rest time was over, and the team moved forward. In keeping with Plan B, they made a brief stop every fifteen minutes, sampled the rocks, and entered the results in our work sheets. After an hour, the stream turned into a series of waterfalls, with small lakes in-between. The path we were climbing now had a steep slope, about twenty degrees. Fortunately, there were only a few trees and bushes to get in the way. Finally, we reached the top. We were quite tired, almost exhausted, so we were very careful about peering over the edge, looking into the crater. Interesting, the view looking down on the inside was about the same as the outside; except the center part of the bottom of the crater was covered with a dense growth of trees. "I'd sure like to go down there and sample the rocks," said a member of the team.

Norm answered, “Good idea. But we are out of time for today. I’ll see if we can put that on our agenda for tomorrow morning.”

We heard the buzz-craft just a minute or two later. We all waved, and Larry “wiggled” his bird to let us know he saw us. A minute later he was on the ground, next to us, and Larry said exactly what we expected, “Load your tools and samples in the trunk, jump in, and we are out of here. We can talk on the trip back.”

Both teams are now in the buzz-craft, and we all want to tell the others about our progress. The Team 1 members gave a good description of what they saw when they looked down into the crater.

Team 2 members were quick to speak, “We expected the same at the top of the hill we just climbed. We didn’t get to see much. It was full of water. We had nothing to look at except this pretty lake.”

“Any fish in there?”

Melvin commented, “Could be, we should try it sometime.”

Norm said, “Wait a minute. This place is two thousand miles away from home. We only come here when we have Larry to take us in his buzz-craft. There are some nice lakes back home just twenty miles away.”

"Yeah right. But what will things look like one or two generations from now. We might see the beginnings of civilization all over Alpha-1. It might even get a little crowded up here."

The conversation was interesting. Then as we approached our destination, we were silent while watching Larry make the approach and landing. No doubt about it. He is a good pilot. As we were stepping out of the buzz-craft, Larry said, "I'll be back with teams two and three in an hour. Then we can have a brief wrap-up meeting. I promise it will not be more than a half-hour, and then we're done for the day."

All of us in Teams One and Two grumbled a bit about being worn out from the day's events. But surprisingly, Norm and Melvin seemed to be keeping up with the youngsters OK, in spite of being double or triple the age of the recent college graduates. The two teams had bragged enough during the buzz flight.

Now they gathered the two teams at separate tables for organizing their notes. They started to lay out their results by types of finds, such as energy (coal and oil), metals (iron, gold, copper, silver, lead, and some rare earths), and other minerals. After a few minutes, Norm said to his team, "You know, this makes for a cumbersome organization. How about grouping by phases of the search, such as river in the flat land, river in the hills, mountain side, and mountain top?"

The team members thought this approach might be easier to organize, and most of them sided with Norm. Melvin, the senior geologist, listened for a couple of minutes, then he commented, "No, I think you are kidding yourselves. Both systems are going to be cumbersome. Organizing by type will be more useful to the geologists back at headquarters."

Norm shook his head, "Organizing by type will take us twice as long as organizing by area. I recommend going for the second choice. Let's sort by area."

The young guys on the team cheered. One said, "Thank you Norm. That will make our job a lot easier, and, also, you can send your report sooner."

Melvin turned toward Norm and gave him a wink, "OK. You win. We will sort by area."

Norm said, "OK gang, the other two teams will be here in another hour. Let's prepare an outline of your report. We'll give those to Melvin, and then he can talk to the combined group. Then our teams are finished for today. Teams 3 and 4 can stay for another hour or two to prepare their draft reports."

Melvin was pleased that Norm was giving him the lead for a part of this very important report. Melvin had searched through the ancient network that was "new" when the ship launched more than four hundred years ago. He didn't really know why he had the urge to see this old information, but he found that it gave him a better feeling of why he is here, as well as a hope that his contributions would be useful for his ancestors in preparing for the future, and whatever else lies ahead.

Larry picked up Norm and Melvin at the appointed time in the morning. As they arrived on the 10th floor of SC253, they were pleased to see Jim and John had the worktables set up, as well as another table with a large pot of coffee and other snacks and refreshments. It was not a difficult job. The four reports were prepared in similar formats, and the only remaining tasks were to create a summary of the four reports and a brief cover letter.

Melvin and John were both experienced geologists, and they made short work of the summary. Larry and Norm worked together on the cover letter. Everything was now in digital format and Larry only needed a few minutes to bring up everything on his computer, select the NSEA HQ address, and tap the send button.

Larry gave a big smile, and said, "Let's go up to 11 and have a game of Pool. We will play teams, my ship against yours. John doesn't play, so that will be Jim and me against Norm and Melvin. OK?"

Norm and Melvin looked at each other, heads nodded, and then Norm said, "Yes, you're on."

On the elevator for a short ride down, everyone was bragging about how good they were. They all knew they were quite good, and they each assumed this would be an easy game. What they didn't realize was the other team was just as good as theirs. This would be a good game.

Jim put the fifteen balls and the cue ball on the table, and pointed to the rack with the cue sticks, then said, "OK gentlemen, pick your weapons."

They did that, then took turns tapping the cue ball back and forth for a few minutes.

Then Larry announced, "Times' up, visitors first," and nodded toward Norm.

Norm answered, "Thank you. Melvin, you can break."

Melvin did a very professional-looking break, with balls evenly distributed around the table. Nothing went in, and Melvin stepped away from the table.

Jim smiled and said, "You left me a number of easy shots. I think I just might run the table." He made four easy shots but made a bad decision on the fifth. He had judged this as an easy shot, then as he was taking aim, he noticed his target ball just might "kiss" another on the way to the pocket. He mumbled, "Well, that won't work," then made a hasty change for another target. That would not work either, so he decided to take what he thought was an "easy" combination shot.

Norm walked around the table and commented while considering his alternatives, “You know the saying, amateurs should never pick a combination shot if there is another makeable straight-in.”

Jim answered, “Yeah, I know, but I’m not an amateur.”

Norm said, “So you say. Now, watch this. Two at a time.” It was a kiss to a ball next to the side pocket then ricochet down to the ball sitting pretty at the corner pocket. Those were clean shots. Then Norm sank five more “easy” shots.

Jim said, “This might be a longer game than I expected.”

Norm’s next shot looked impossible. His target was sitting in front of the side pocket. There were two other balls in the way, so a “direct” hit was out of the question. “Here’s what I’m going to do. I’m lined up for a ricochet off this cushion, then across the table and sink my target into the side pocket.” It was close. He showed a sad face and stepped away.

Larry had been studying the remaining possible shots. Six targets remaining on the table, and the 8-ball has to be last. Larry decided to follow Norm’s style and “think out loud.”

“Well, I don’t see any easy shots. Hmmm, except this combination looks easy enough; and that will put the cue ball in a much better position for my next shot." Then he very quietly took his position, aimed, and made his shot. "There. Two down, and four to go." The next shot looked impossible, with the remaining target in a direct line behind the eight ball. "Behind the eight ball. Well, watch this." The cue ball ricocheted from the cushion to send the target to a corner, and the cue ball a long way away. The eight ball was not touched. "Melvin, you have a lot of green between your cue ball and the target. Good luck."

Melvin took a minute to study the layout. "Actually, this looks pretty easy." It was a clean shot, almost dead center on the target, then off the cushion and to the far corner … and in the pocket.

Norm spoke up, "Nice shot. Only one to go."

Melvin had already worked this out in his head. "And now for the easiest shot of the day." With the cue ball and the eight ball both near the middle, and only a foot or so apart, it was an easy shot. Oops, the cue ball rolled to the corner pocket and dropped in. Poor Melvin hung his head, and he looked like he could cry.

We all applauded. We enjoyed the game and we wanted to play another.

Larry gave the signal to quiet down, "Look, I have some administrative stuff I need to take care of this afternoon. I would like for this group to join me tomorrow morning to start our planning for another geologic search."

The group let out a "groan" to indicate they were not thrilled about Larry's plan.

Larry continued, "There's no need to wear ourselves out. Norm and Melvin, can I pick you up at eight-forty-five in the morning?" They both nodded in the affirmative.

"Great. We will keep this between the five of us until later. Think of it this way. We will do the planning, and the young guys can do the work."

Norm spoke, "Way to go, Larry. I like the way you think."

Larry continued, "Thank you. Now back to business. Norm and Melvin - let's go down to the bird, and we will buzz out of here. Jim and John, while I'm taking care of my administrative duties, the two of you can set up the equipment this afternoon for what we will need tomorrow morning. So far, we have only looked to the east. Let's look to the west this time, and let's try to keep our searching a little closer to home. Say, within two hundred miles."

Day 7. The next morning, Norm, Melvin, Larry, Jim, and John were gathered on the 10th floor and ready to go to work. Larry waited for everyone to fill their coffee cups, then took the lead, "Well, I am not a geologist, so Melvin and John will take the lead after I walk you through some logistics. Earlier, we had good reason to do our search two thousand miles away. Today we don't, so let's stay close, like not more than a hundred miles. With the help of the buzz-craft, we can keep our flights under ten minutes a trip, and that makes those logistics a lot easier."

The others all nodded to indicate their agreement.

Larry continued, "Let's plan on five teams. The five of us will be team leaders, and then we can have one more on each team from our group of recent college graduates. With our sidecar we can carry ten people, and that removes the need for multiple trips. Any comments, or questions?"

Our planning meeting continued into the afternoon. Larry said, "I believe we are nearly done for today. I am going to take Norm and Melvin home. Jim and John, the two of you can go review the list of recent college grads and pick out five of these guys to join us for tomorrow's searches."

Day 8. As planned, Larry picked up Norm and Melvin at eight am. They joined the rest of the team on level 10 for a brief review of today's plans, then everyone donned their work clothes, picked up their tools, and joined at the buzz-craft, sitting in front of the ship.

In the air at 8:40, we are in a good mood. Then conversations quickly ended as we realized this landscape was different from what we had seen on earlier flights. Larry noticed the sudden quiet and he held the speed down around 400 mph.

Our “target” for this morning is a mountain - actually an inactive volcano – sitting out here in the middle of the water. As we were landing, we realized there was nothing else that could be seen. Just water - and this mountain. Nothing else to interrupt the horizon. There was a small beach area at the water level. Melvin told us the molten lava tends to solidify there and adds to the island a small beach, sometimes all the way around the island.

For this site, we are positioning our five teams on the beach, each separated by about 50 yards, with some simple directions. Using a slow to medium walk, they head toward the mountain peak. We operate individually, stopping and recording samples as appropriate. At the end of two hours, Melvin blew his whistle. That is our signal to stop. Then we take a 15-minute break. Then we face the water, step to the right about 25 yards – that should be about halfway to the path just completed by your neighboring team. Then we take a new heading toward the beach and collect more samples.

Time is passing pretty fast. It is now twelve-thirty, and we took a break for lunch. This was a good time to relax, and we expected this would be our brag-time; but we soon found we didn’t have much to brag about. Most of what we saw was repeats of what we found in our earlier search. The only thing new was some evidence of oil in several of the areas. The most common clue for oil was a black "smear" on the rocks. Melvin helped us understand. "Just imagine that these rocks were in a pool of oil several millennia ago, and slowly the oil disappears, leaving this black smear. Formed in this manner, the black smear appears to be part of the rock. If we were looking for oil, we would start digging around this area, trying to find where the oil went."

Several in the group got a little excited, "Can we start digging now? Do you suppose we will receive a bonus if we strike oil?"

"You might want to take a look at one or two shovels-full of dirt, just to see if there is any evidence of "wet" oil. Other than that, just make a note of the location so the oil-hunters can find this spot - if they are interested. This is a good find, and you will most likely be thanked."

For our next assignment, we loaded everything into the buzz-craft, and flew around to the opposite side of the mountain. Larry found a reasonable spot further up the slope. We quickly spread out to the left and right, uphill and downhill, and all of us within 50 yards. Larry warned everyone to be careful. "We don't want to lose anybody tumbling down the slope."

After an hour, Melvin blew his whistle. That was our signal to end the search and return to the buzz-craft. We loaded everything, and Larry headed to our next site, just 40 miles away. As we approached a long coastline, we could see the terrain was completely different. Everything up and down the coast appeared to be flat land, and about two or three miles inland we could see a long row of mountains. That is where we're headed. Two or three minutes later, Larry slowed and started looking for a place to land. He said to Melvin, "How about around the base of this mountain?"

Melvin shook his head and pointed upward. "No, let's fly up above the mountain tops and see what things look like from up there. I would really like to find some recent volcanic activity."

Larry took us for a slow ride, just a half-mile above the mountains. After five or six of these, Melvin spotted what he was looking for. "There. See the flat top? Let's take a closer look there."

A minute later they were looking down into the crater, about 50 yards across. "That's what I was looking for. Larry, can you take us down in there?"

"I thought you would ask for that Well, the answer is no. See the trees down there? We really would not have a place to land. What we can do, though, is land on the rim. It looks like we have plenty of room for that." Melvin nodded his head. Larry circled one more time and landed. Melvin identified five locations for the other team members to look for interesting samples.

Larry approached Melvin, "How would you feel about going down there on a rope?"

Melvin was surprised, "Yes! I'll do anything to take a look down there. What do you have in mind?

"We keep a long rope in the back trunk of the buzz-craft. It has come in handy several times for lifting things out of awkward spaces. I think it would work here. We also have a harness that you will want to put on. This is similar to the harness that is worn by military people for jumping out of airplanes with a parachute on their back. The cords from their parachute are connected to the harness. Same set-up that we have for you."

Larry motioned for Norm to join, then brought him up to date. "And we will need one more person to make this work."

Norm looked surprised, "And what is that?"

Larry continued, "We don't have a windlass for reeling the rope in and out, so I will be flying the bird up and down, with Melvin at the end of the rope, The problem is I cannot see straight down. The floorboards are in the way. So, I will need for you to lean out your side window and give me directions." They talked about how best to do this, and they were ready.

Using the rope, they measured the distance from the ground down to the bottom of the crater. "Forty-three feet."

Norm pulled out his phone camera and focused on a small bush at the bottom. "Forty-two and a half. Close enough." Larry landed the buzz-craft, marked the rope at forty-five-feet and fastened the rope to a rigid place in the trunk.

Melvin was in his harness. The rope had a snap-hook positioned just over his head, easy for him to reach with both hands, in case of an emergency. Larry held up his hand in the thumbs up manner, as a question. Norm and Melvin answered "yes" with their thumbs up. Melvin tapped the "up" button and started the buzz-craft moving upward. Slowly. When there was only one more coil of rope on the ground, Norm gave a slow-down motion with his hands. Larry reduced the vertical speed by one-half. Melvin gave a thumbs up when he was lifted off the ground. Norm relayed the thumbs up to Larry, and then Larry also showed thumbs up, and stopped the climb. With Melvin about three feet above ground, Larry slowly moved the buzz-craft over, so they were about ten feet inside the crater.

Norm was carefully watching Melvin and saw him moving his head around, looking at the new view. Then Melvin looked up and saw that Norm was watching. Melvin wanted to talk to them. Talking out loud, "But yelling to the top of this hole won't work.

If I only had my phone . . .But I do have my phone! Melvin started fumbling through his many pockets and he found it. Norm was surprised to hear his phone ring. He answered, “Hello.”

“Hi Norm, we should have remembered the phone earlier. Can you hear me OK?”

“Hi Melvin, I hear you loud and clear. I’m putting you on the speaker so Larry can hear.”

Melvin continued, “OK. Now that I’m down here, what’s the plan? I’d like to spend some time here looking around, and then collect some samples.”

Larry answered. “Melvin, how do you feel about disconnecting from the rope. You can look around for an hour or so, then I’ll come back, and we’ll drop the rope to you, Will that work for you?”

“Perfect. That’s exactly what I wanted. But I have one condition.”

Norm answered, “Oh, and what’s that?”

“I want you guys to promise you won’t run off and leave me. It is a little spooky down here.”

They agreed to one hour, and Melvin would ring Larry’s number if he wanted an earlier pick-up. That done, Larry parked the buzz-craft near the crater, then coiled up the rope and locked it in the rear storage area. Larry had planned on visiting with some of the active geologists, but then decided to stay here to keep everything safe. What if something happened to the rope. Melvin could be trapped at the bottom of the crater. Melvin's imagination went to work. "If I were trapped down here, how would I die? I could starve to death. Or I could break my neck trying to climb out. Or I could be slowly chewed up by a swarm of ants, or other critters." He felt a cold shiver going up and down his back. Enough of this.

"Norm, any word from Melvin? How's he doing? How are the guys doing? Anything interesting?"

Norm was still twenty feet away, "Wait one. I'll be there in a minute. Is everything okay? You sounded anxious."

Larry had calmed down, "I wonder how Melvin's doing. Have you heard anything?"

Norm answered, "Yes, he called me about five minutes ago. He said this place is rich with about every mineral you can think of; and especially rich in metals, including gold, silver, iron, nickel, and copper. He said he has seen enough, and he would like to come up as soon as we can make it happen.

"Larry said, That's fine with me. Give him a call. Tell him we'll be on the way in a few minutes."

Norm helped retrieve the forty-five feet of rope from the rear storage compartment. He attached a small bag containing about 5 pounds of dirt. No rocks. Norm asked, "Can we just toss this dirt-bag into the crater? Will it unravel?"

Larry said, “No, I’m afraid it might be tangled with bushes on the way. Jump in, and let’s get on with it.” After Larry saw Norm’s thumbs-up, he added some power to climb the buzz-craft up to 50 feet. When Norm saw everything was clear, he gave Larry another ‘thumbs up’.

Then Larry moved the bird over, so the dirt-bag was hanging directly above where they lowered Melvin earlier. Norm gave another thumbs-up and then started down. Slowly. When the dirt-bag was a foot or two above the upper ground level, Norm made several adjustments so the dirt-bag would have a clear trip all the way down. They started down again, slowly.

Mel spoke to Norm over the phone, “Hey Norm, what’s in the bag? Is that my lunch?”

Norm and Larry both laughed. Then Norm answered Melvin, “You are funny. That’s just a little weight to help keep the rope from tangling.”

“Got it. Thanks. Is it OK for me to disconnect the dirt-bag?”

Norm answered, “Yes. Oh, I should have explained this earlier. “Do you have any samples to bring out?”

“Yes, I do. Probably 40 to 50 pounds.”

“OK. Dump out the dirt and put your samples in the bag. Then fasten the bag to the lower cord on your harness. And then connect the upper cord to the snap at the end of the rope, and you are done. How are you doing?"

"All done. I'm ready to go up. Please."

With Melvin above the rim, they positioned him away from the hole. Finally, standing on solid ground, Melvin unsnapped his connections. Larry parked the buzz-craft nearby, then he and Norm started coiling the rope. Melvin walked over and commented to both Larry and Norm, "I was purposely quiet regarding the samples while I was down there, because I didn't want to get too excited. But now I can tell you, this is the richest set of any of the samples I have read about. I found samples of just about everything I can think of, and at least double the normal density. I would expect NSEA will find a way to share in the profits from this. Maybe we can be considered as employees and then we might receive a nice bonus. Larry, you are from the recent times, what do you think about that?"

Larry had a serious look on his face and said, "I don't have any experience with those things, but I will certainly make a recommendation. You guys be patient, and I'll keep you posted. Now, let's see how the other guys are doing."

John Collins stepped forward, "We thought maybe you forgot about us. Here's my summary. This area includes the same minerals we saw this morning, plus a few more – but the density is about double. We also found heavy traces of most of the metals. And we also found slight traces of coal and oil. In short, there is a lot of everything. How's that?"

They were all quite happy with these results, and they headed for home.

Back at Larry's space cruiser, they gathered their samples and moved to level 10. John had already talked to the kitchen crew and arranged for two pots of coffee, a pile of snacks, and several desserts.

Their first goal was to create a format for the next report to NSEA. Larry asked for volunteers to work on various sections of the report. Melvin started to school them on a reasonable structure for their reports, Then he turned to face Norm and Larry and said, "You know, these guys have had a long day. I believe it is time to call this a day and continue in the morning."

Larry stood up, "Thanks Mel, I agree with you. We are done for today. And then I would like to see all of us at ten o'clock in the morning."

Day 9. The next morning, Larry had planned on a short day to finish the second report to NSEA. But he found waiting for him the NSEA response to his first report. His first task for today, then, is to call Norm and Melvin,

Larry phoned Norm and Melvin to tell them about the letter from NSEA, “So, I need both of you to help me this morning. NSEA wants us to use a different format for our report. When that is done, I’ll tell you about our new tasks. Can I pick you up at eight?”

On the trip back to SC253, “Melvin, remember for our second report you recommended including a description of the difficulty in reaching the products in the various locations. Well, we should have done that in the first report. We need to review the early team reports and add for each location a brief discussion of any difficulty. Also, include one or two pictures for each location. Are you OK with that?”

Melvin answered, “Yes. That should be an easy task.”

“Norm, it appears the content of our second report is about right. But it is still in the draft stage. So, I would like you to take half of the report and do a good edit. Make sure we provide detailed information the way they want it; and include pictures. I will do the other half.”

Norm answered. “Larry, I am happy to help. You know, this might be an all-day job. Are you providing lunch?”

Larry laughed, “Yes, John is already at work, setting up our work tables, and two large coffee pots. In a few minutes he will bring in snacks, and later on we will take a break for a nice lunch.”

On the 10th floor, everything was ready. There was one table for Melvin, with all of the original notes, draft reports, pictures, and the final report. Another table was for Larry and Norm, with everything they needed to prepare the final version. The third table in the room contained two coffee pots and other related paraphernalia.

Conversations on the 10th floor this morning were quiet, and task related. John showed up pushing a cart loaded with orange juice, sweet rolls, cookies, and other snacks. They must have been hungry. Norm and Melvin stood up, studying the layout and deciding where to start. Conversations quickly moved to the snacks, then migrated back to comparisons of the general terrain of Earth to Alpha-1 and, briefly, hunting, fishing, and golf. And then the room was quiet again as the three experts returned to their tasks.

Two hours later, Melvin was finished. He moved to the other table, and saw they still had a way to go. Melvin was the only expert geologist in the room, and he wanted to help, "I think you need some help," and he moved what amounted to a third of the remaining total in front of himself, and said, "I'll do these."

Larry replied, "Thanks."

Another hour, and they were finished. John had been monitoring their progress, and wheeled in the lunch cart, "Lunch is being served." Jim Womack, the Operations Officer for Larry's space cruiser, was following John, and said, "John looked like he had his hands full, so today I am John's assistant."

"Great timing, John – and Jim," said Larry, while he was patting John on the back. Then he gave an exaggerated wink to Jim.

John and Jim were setting up a table in the room next door. "Away from all the work-stuff," John said to the group.

Half an hour later, Larry cleared his throat and said, "I want to thank everyone here for your help, not only for today, but for the past two weeks as well. In my letter to NSEA headquarters, I have described your assistance. Now, I need your help on an item. We had a short conversation yesterday about some sort of reward for our work here in finding such a treasure for NSEA. We all agree that NSEA will probably make a fortune from this. We would like to have a share of that. Three of us, John, Jim, and myself, are NSEA employees, so it is easy for NSEA to issue us a bonus for a good job. Norm and Melvin, however, are not employees. They have performed in leadership positions, and I have no way of compensating them for their time, or for a bonus."

Then Larry paused, hoping someone would have an idea. The others were silent. "Well, I have thought about this quite a bit. On my first day here, just two weeks ago, I told Norm and Melvin, both, I would like to have them help out on our geology search. I needed Norm because of his operations talent. No one knows this planet better than Norm. No one is better at making decisions in complicated situations. I needed Melvin because he is a geology expert. No one knows that subject better than Melvin."

"OK, so I made a good decision in bringing them to our team. But shame on me for not completing some paperwork to make it official."

Larry's original plans for the report started with specifics of the products, what we found, where we found it, and some mention of the quantity. This section should match the format that Melvin suggested. The second section would give a day-by-day summary of what we did, and the third would discuss who did what."

We had a lengthy discussion. Larry was worried for a while, until he realized the main difference was merely the order of the three sections. The new plan starts with an overview which will be a summary of what we did and will mention the roles of the key personnel. The second section will have sub-sections describing in detail just how the leaders and managers contributed to the success of our mission. The third section presents the details on the products we found, and comments on the ease of reaching the various areas.

Finally, they all agreed that this was the proper format for our report. That settled, we needed only an hour to rearrange the sections of our text to match our new format. With that done, Larry assigned Jim and John to do a complete edit and then send it to NSEA.

Larry took Norm and Melvin to another room to brief them on our new tasks from NSEA, "First there was a paragraph saying they noted we were fortunate to find some talented individuals to help with our tasks, and they hoped we would provide more information in that regard." Norm smiled. Melvin said, "That sounds like payday."

Larry commented, "Congratulations to both of you. And there is more. Both of you are now NSEA employees, effective the day after I landed SC153 on Alpha-1 and continuing for at least another month. Now, let's look at our new tasks. NSEA wants us to stay where we are for perhaps another two weeks. In the meantime, they will be trying to figure out what to do with us for the longer term."

"So, our primary task is to continue gathering samples. They would like us to put the new college graduates to work, with us older employees doing the planning and the young guys doing the actual work." Larry stopped for a moment, then, "I think we can agree to that. Right?" Norm and Melvin both smiled and gave their thumbs-up signal.

Larry continued, “They also gave us a simple schedule. I will read a paragraph from their letter.” Reading from the letter, “You will divide the grads into four teams. Odd-numbered teams will be used on odd-numbered days, and so on. We know this is shared work, so teams will work on-site for four hours a day. This is tough work, so team members will still be paid for eight-hour days.”

“Their schedule suggests we plan on four trips each week, and one report at the end of each week.”

“And that is the plan. Their letter suggests they will let us know the overall plan later. I’m sure you are as anxious as I am to know when that will be, but I think we should just be patient. “

Larry sent a message to all concerned, giving them a very brief version of our new tasks, and telling them to join us at level 10 for the details at nine tomorrow morning. There was a separate note to Norm and Melvin that he would pick them up at 8:30.

A new routine.

Day 9, Jim and John were in the meeting room at 8:00 to arrange tables and chairs and set up a separate table for coffee and some other delights. The head table was reserved for Larry, Norm, and Melvin. Jim and John had chairs at the coffee table.

At nine o’clock sharp, Larry opened the meeting, “Good morning. By now, I’m sure everyone has taken a few minutes to review the message I sent yesterday. The people at NSEA Headquarters did a good job with the schedule they sent us. I wish I had thought of something like this earlier, so we could have brought more of the recent college graduates into our geology searches. With this approach, all of you will be going out for two hunts each week. Now, you are probably ready to ask me how long this might last. Well, I don't have a definite answer, but I'm fairly certain it will be at least a month."

Larry continued telling the geology grads what he knew, as well as some guesses at what might be ahead of us. There were plenty of questions, some of which Larry promised to forward to NSEA for answers,

After a short break for lunch, John took over the chore of signing up the 22 grads for their team assignments. Larry will be making one trip for each of the four days each week. The buzz-craft will carry only nine team members, leaving four grads on standby each day. This looked like a difficult math problem, and several in the audience were arguing which method would be the best. John hollered, "Stop! I will take care of that, so you won't have to do anything."

The group quieted down, and Larry took charge of the meeting. "We're about done here. Not more than five minutes. Your geology trips are based on half-day searches. When you finish a search and are back in the ship, John will provide you with a blank form we want you to fill out and provide back to him. These reports should not take more than a half-hour. On Friday, we will gather all of your reports, combine the information, and prepare a summary of our wok during the week. This task will be performed by Melvin, Norm, John, and myself. OK, now let's talk about tomorrow.

"Tomorrow is an odd-numbered day, so we will be using teams 1 and 3. For tomorrow, the first member of Team 1will be on standby. The next time Team 1 is on the schedule member number two will be on standby. After all members of Team 1 have been on standby, Team 3 will take that duty."

"We will gather here, on Level 10, at eight in the morning. Team members will be issued work clothes and lunch bags. Then we move to the ground level, load the tool carts in the buzz-craft, and then we'll be airborne at about nine. We will head out in the northwest direction. We are only going out to two hundred miles, so our flight time will be about half an hour. The spot where we land the buzz-craft will be the starting point for both teams. Team 1 will start their first heading to the northwest, then stay on that direction for one hour. During that hour, your routine should be to walk ten minutes, then look for samples for another ten. At that rate you will search three small areas. Then turn to the southwest for the second hour and follow the same routine. After this second stretch the clock should be about noon, so this might be a good time for lunch. The third stretch should be heading southeast for an hour and then one hour to the northeast should return you back to your starting point. Team 3 will be on a similar route, but east of point zero instead of west. The two teams should be back at point zero about two o'clock or two-thirty. Our return trip will be another half hour. Then you can shower, change clothes, complete your report, and you're done at about four. That is a full eight-hour day."

Larry had finally come to the end of his list, "Well, that's all I have on my schedule. I certainly have time for a few questions. Who wants to be first?" Silence. "No questions? Great. This meeting is over, and I will be here with Teams 1 and 3 tomorrow morning at eight o'clock sharp."

Larry motioned for John to remain, then pointed to a couple of chairs. John spoke first, "What's up Larry?"

"John, we first met when you were the senior geologist on our trip to Pluto, two years ago. Tell me, what did you do before that?"

"Well, I graduated with a BS in geology seven years ago. I was surprised that jobs were easy to find. Most companies would hire a team of temps for a big job, then lay them off as soon as that was finished. One of my temporary jobs was with NSEA for a three-day trip to the moon. That was an exciting time, and then I went job-hunting again. A couple of months later the NSEA recruiter contacted me for the Mars trip. We were only on Mars for two days, but the whole trip was three years. Space ships moved a little slower back then. After that trip the recruiter signed me up for a series of jobs to other locations, some of these to look for important minerals and some to survey land features. The trip I made with you to Pluto as a senior geologist was even more exciting,

"I was wondering if you ever had a desire to go back to looking for rocks."

"Ahh, no. It is an interesting occupation, but what I am doing now is more interesting.

Larry continued, "OK, then, let's talk about what you and I need to do for our mission; that is, for the next month."

Larry described that his earlier thoughts were focused on how long it would take to check out a set of sites that could be representative of the entire planet. With a little research he learned that the surface area of Earth, in round numbers, is about twenty billion square miles. The circumference of Alpha-1 is about ten percent less than Earth's circumference, so the surface area of Alpha-1 should be about eighteen billion square miles.

"Now, let's try sites of ten thousand square miles each. That would be one hundred by one hundred. After some not-so-easy math, I computed we would have to search more than eighteen billion sites.

At this point, John spoke up, "I can't dispute any of your calculations, but remember that half of the surface area is covered by water, so we can drop that to nine billion sites."

"Thank you. Every little bit helps. Now, pretend we have a thousand teams, and we are going to search every day of the year. We still need twenty-seven thousand years to complete the search."

John spoke up again, I think we need to consider a completely different approach. How about we go back to our original approach and just do what we can do. They didn't tell us to search the whole planet. They told us to use the twenty two fresh graduates, and search five days a week. Why don't we just do that?"

"Thank you. That was my original thought, but I didn't want NSEA to think we took the easy way out. Or maybe you had some other ideas that might be worthy of consideration."

John smiled and said, "Okay, we're taking the easy out. I have already worked out a basic approach for this. Where do you keep the maps, and then let's get started."

Larry needed only a couple of minutes to turn on the viewer on one of the work tables and pick a map that included the last area they had covered a few days ago. The one about a hundred miles to the west. He placed a finger on the spot and said, "Let's plan on taking Team A to a spot fifty miles north of there. Then I will cycle back for Team B and take them fifty miles further to the north, and then back for Team C to go another fifty miles north.

Chapter 9 – Recall to NSEA Headquarters

Two weeks later Larry received an important message from NSEA. They urgently wanted him to return to NSEA ASAP. "You are directed to schedule lift-off not later than the day after tomorrow. You will return Space Cruiser SC253, and all of the 120 crew members. The heavy-duty equipment can be left behind as a gift to the managers of Our Town and West Village. This includes the small combination bulldozer and fork-lift, and also the small dump truck. Also, please invite the following to join for the trip: from SC12, Norman Link and Melvin Sherwood; from SC16, Dave Parker: and from SC15, Maynard Jones. NSEA management would like for those invited to stay for at least three days, and we would be happy to extend their stay as long as three weeks if they so desire."

"Your flight time will be twenty days, so you may want to include some of your favorite choices in the kitchen. There will also be plenty of synthetic meals available from our standard menus. They are adequate for physical health purposes, but I'm sure you will remember the taste is not first-class."

Larry forwarded the NSEA message to all concerned. His message included, "Today is Monday, and I am planning for lift-off at ten AM Wednesday. For those invited to join the trip, I will need your decision today. If your answer is yes, I can pick you up Tuesday afternoon. I suggest you bring four or five days of your personal clothing, all washable. I can provide you with the normal attire we use for in-flight periods.

Larry phoned Maynard Jones. Yes, he is delighted to join us for the trip. Three days will be plenty. For the heavy-duty equipment, Larry will have them outside, on the grass, later this afternoon. Larry will pick up Maynard and four technical guys eight o'clock Tuesday morning to start moving the equipment to Our Town. The distance is about ten miles, and they thought they might need only one day.

Larry phoned Dave Parker. He is looking forward to the trip, and he might be interested in a longer visit. Larry will pick up Dave at two PM tomorrow.

The call to Norm and Melvin went well. Norm said "Yes – No - Maybe. Look, I am still a bit undecided about returning to Earth. I'll call you back later today." Their pickup was tentatively set for five PM tomorrow. Melvin said, "Count me in for the three-day visit, regardless of what this other guy says."

It was an exciting evening, with everyone happy about the unexpected trip. The twenty-two recent graduates all looked forward to going home. Eventually, Norm made the call to Larry and said, “Sorry I was so undecided. I will be happy to join you for the three-day visit.” Larry answered, “Great. I’ll be picking you up at five tomorrow afternoon,”

The helicopter arrived at eight sharp, with Maynard and four technicians. Maynard brought a fully loaded suitcase. The techies looked over the equipment and started the engines. “We were looking at the many trees on the way in and saw a lot of trees that will need to be removed. This will be a two or three-day project. It’s a good thing we happen to have this heavy equipment to help us.” Larry escorted Maynard and his suitcase to his bunker. “We are a little short on space, so I hope you don’t mind sharing the space with Dave Parker.” Maynard replied, “That’s OK with me, as long as he doesn’t snore.”

Larry picked a couple of the recent graduates to join him for a walk around the area. The NSEA rules require that he carefully look over the area out to a hundred yards and remove any burnable material. Most important, they did not want any fires started that might transfer into a forest fire.

Norm and Melvin and their two wives were sitting in chairs on their side-porch when Larry parked the buzz-craft in Melvin’s front yard. Norm motioned for Larry to join them, then he said, “We are not allowed to make the trip until you agree to the conditions our ladies have imposed. Melvin and I have agreed, and they want to verify that you will also agree."

Alice, Melvin's wife, stood and said, "I only have one request, and that is to return Melvin ASAP after your three-day visit."

Larry replied, "I will do my best to return Melvin promptly after our visit,"

Janice was already standing, and now she looked directly at Larry with an authoritative expression. "Larry, I'm sure you will return Norm on time. I also want you to protect him from harm so he will be able to return to me in the same condition he is in now."

Larry gave a big smile, then answered, "I promise to do all I can to provide protection to both Melvin and Norm."

Following a brief conversation and their final goodbyes, Norm and Melvin loaded their suitcases and they were off.

All who were expected are now on board. Larry had a short list of things to cover, and he invited everyone to tune in on their AVCs. "Hello everyone, this is Larry, and I have a couple of things to go over. The first thing is to inform you the doors are locked. We just performed a complete head count, and we are all here. If anyone has a need, let me know what the problem is, and we will do what is necessary. In the event of a fire or some other emergency, the doors will automatically unlock, so don't worry about that. New subject. Twenty days can be a long time, and boredom can make it seem longer. Your AVCs provide access to a long list of books, movies, and computer games. You will also find pool, ping-pong and darts on floor eleven."

"New subject. As I told you earlier, lift-off is scheduled for ten in the morning. I need everyone to be in your assigned spaces by nine o'clock. You are free to move around for a while, and then strapped down by nine-forty-five. The lift-off itself is interesting to watch, and it is even more so when you are a part of this. You will have to remain strapped down for at least an hour after lift-off. We have a series of accelerations to go through, and I will give you ten-minute alerts for each of these. That's all I have for now."

"Oh, wait. One more thing. You don't want any loose items in your bunkers while we go through the lift-off and acceleration periods. I ask you to take advantage of that quiet time starting at nine in the morning and put things in a safe place. Your bunkers have several drawers with latches, and they are very safe. OK, that is all for now. We will talk again in the morning. Have a good night's sleep."

Things were going well in the morning, until Larry's warning light on his AVC turned on. He picked up his 1MC microphone, "My AVC is showing signals that say two people on the twelfth floor are not strapped in. I need for you to strap in NOW! If not, I'll have to send someone to take care of this, and that might cause a delay in our lift-off. I told you yesterday this was important. Remember, I am not allowed to lift off if anyone is not strapped in." Larry saw the two lights turned off, and he said, "Thank you, now all lights are off. Lift-off will be in five minutes."

Larry gave the signal to Engineering for engine warm-up. The soft, growling sound caught everyone's attention. A minute later Larry signaled for 25%. The sound was louder and at a higher pitch. Another minute, more power, everyone felt the ship move. More power, then they felt the ship starting to move upward. Then faster… and faster. The acceleration was constant. Larry spoke to the crew, "We are currently doubling our speed every fifteen seconds. The acceleration will cut back in another minute or two. And then we will do our first jump."

The acceleration was off, so now we felt like we were floating. Larry said, "Now listen up. This is the important part. We will have three jumps. You want to be on your POD, sitting up or leaning back, facing the pointed end of the ship, with both straps fairly tight. The first jump will start in about a minute. You won't be able to move for five minutes." A minute later, "Here we go. Hang on,"

Five minutes later, all is quiet. We can float. Larry said, "That was a good one. The gauge here indicates we are going eighteen times the speed of light. We will do two more jumps, and that should put us around thirty times the speed of light. We abbreviate that as 30L. We will take a ten-minute break now, and then a four-minute jump. And then another break and our third jump. And then we can relax for a while."

"In order to land our ship at NSEA, we will have to go through this routine again to slow down to our approach speed. The first step involves turning the ship around. Since we are just coasting, that is an easy task. Besides, the computer really does the work. Then, as the thrusters are activated, they will slow us down instead of speeding us up.

Before long, we were at cruising speed – about 30L. Thrusters are now off, and everything is quiet. Norm and Melvin are roommates, and they had spent a good bit of time trying to figure out why the managers at NSEA wanted to talk to us. Is it possible they learned that Alpha-1 is rich in the natural resources that are so important to earth. But why talk to us? Maybe to give us an award. "Let's ask Maynard and David to join us."

Norm punched the buttons to ring the AVC in their bunker, "Hi Maynard, are you getting used to the quietness?"

"Well, not yet. It is sort of spooky."

"Melvin and I are in a deep conversation trying to figure out why we have been summoned. We have not found a reasonable answer yet, and we thought we could do a better job if you and Dave could join us. We are in bunker D, one floor above you."

Five minutes later, Norm opened the door and said, "Glad you're here. What took you so long? I was wondering if you were turning down my invite."

Maynard replied, "Sorry we took so long. Dave was sound asleep, and he is difficult to wake. So, what's up?"

"Well, we're trying to figure out what NSEA wants out of us. My thought is they want to make heroes out of us. Heroes for making the trip. Heroes for starting a new civilization. Heroes for helping each other. And maybe heroes for providing geologic information back to them. And maybe a thank you for that part. Do you have any other ideas?"

Maynard answered, "Well, I like that last one. Is it possible that the geology on Alpha-1 is a lot better than on Planet Earth. What do you think Dave?"

In a sleepy voice, Dave commented, "Huh? So, we dig around in the dirt for a few hours, and eventually we come across something useful. I'd say it's all the same. "

Maynard said to David, "Go back to sleep." Then he continued, "It is possible that NSEA realizes

Alpha-1 is rich in the natural resources that are so important to Earth. If they can come up here and dig in the right placcs, they can cover all the shortages. And they don't have to move any workers up here. We already have plenty of workers."

Melvin was anxiously waiting for the right time to say something, "NSEA already knows what's up here. Now they need to provide some heavy equipment, and some small ships that can move stuff here and there. And we will need several launch and recover pads – for the big ships, like on the NSEA campus back on Earth."

Norm finally spoke up, "I really like the way you guys are thinking. But decisions like these will take time. If I were involved, I would want to establish an office on Alpha-1; with several landing pads; and bring up some work ships; something like Larry's bird. They wouldn't need to be so fast, but it would be nice if they could carry twenty or thirty people. Certainly, smaller than the shuttlecraft that can carry one hundred and fifty people."

Melvin chimed in, "Maybe one shuttlecraft might be a good idea. In addition to a couple of the smaller birds."

Maynard had an anxious look on his face, "Shouldn't Larry be in this meeting? He might know the answers to some of our questions."

"Well, I'm glad you brought that up. Larry told me he received a personal message from HQ. The message said they recognized Larry may have developed relationships with the leaders at Alpha-1, and they wanted Larry to remember he is an employee of NSEA. Larry told me he will definitely be on our side, but he intends to avoid any meetings that could appear to be moving him further in our direction."

With no work to do, many of the crew have started visiting the sport facilities. Shortly after their meeting ended, Norm and Melvin stopped by the pool room to see if there was anything interesting going on. Nobody was there.

Norm spoke first, "Oh yes, we are at our cruise speed. A solid break-shot would result in balls all over the room. With zero gravity everything floats. Actually, Larry likes to use what he calls near-cruise speed. That way, people can stand up with their feet on the floor, and with some care, they can actually walk around."

Sometimes it is fun to watch. Many people actually bounce as they walk. Some like to use a toe-push that can put their bounce three or four feet in the air. Some have even bounced back from the ceiling.

Melvin said "Let's check out the basketball court. Maybe we can play 3-D basketball. The basketball court is on this same floor."

One of the young guys was bouncing by, and Melvin approached him, “Hey, how come nobody’s using the basketball court?”

The response was interesting, ”Are you kidding? How can we play basketball when everything floats.”

“Well, I guess twenty-day trips aren’t enough to cause you to use your brain and invent new things. When we had our four-hundred-year trip, my forebears invented a new sport, called 3-D Basketball. Where can I find a basketball?”

The young guy retrieved a basketball from a locker and tossed it to Melvin, “Here you go.”

Melvin said, “Thanks.” He dribbled closer to the net, then jumped up to the net and dropped the ball inside. “Over the years, we prepared a complete set of rules for our game.”

Norm said, “Our ship was equipped with nuclear-powered machinery. No one had any idea how far, or how long, our nuclear seeds would last, so we started off with a very conservative approach. As it turned out, the consumption rate was so minimal, we had difficulty with our measurements. At the end of one year, we recorded it as a minimal consumption. At the end of three years, we saw enough change to make an estimate. Our first estimate was we had expended about 0.04 percent of the total supply. That told us the original supply could last us for about nine or ten thousand years. Our current speed was as fast as we could go and would adequately provide us at Apha-1 after a total flight of four hundred years.

That information told us we didn't have to be too skimpy about our fuel usage. Early in our trip we made notes of the mood of our crew and passengers. It was common knowledge that, after we have finished accelerating to our agreed-upon cruise speed, flying at a constant speed is the most conservative speed. But constant speed puts us in a situation where everything in the ship is flying at the same speed and everything floats. There is absolutely no feeling of gravity, which we call one G. On the other hand, when we needed to slow down, we set the computer to slow us down at a rate of one G. When we are coasting at cruise speed or speeding up, the pointed end of the ship is up front, and we can move around the ship with our feet on the floors. This is the most comfortable situation for all of us. Then, when it is necessary to slow back down, we flip the ship so the pointed end is behind us, and we are again able to walk around with our feet on the floors.

Our new friend eventually shook his head and said, "I don't think I'm interested in your 3-D game. I like regular basketball." The subject was dropped and never came up again.

Melvin turned to Norm and said, "Well, I guess we'll have to get used to the normal ways to pass time, reading, movies, card games, computer games, and napping. "

Norm replied, “Let me know when you get bored. We can watch a movie together. Did you notice the movies here are at least four hundred years newer than the movies we used to watch.”

Landing at NSEA. Early in the trip, Norm asked Larry if he could be a visitor in the cockpit for the approach and landing.

Larry quickly answered. “Certainly. If the computer fails us, I May need some help. I would like to hear how you learned to land. It is obvious you had no flight instruction. Can you walk me through that?”

“Well, someone in the original crew, twenty generations earlier, started a PDL system. That stands for pass down the line. Everyone in an important position was required to create a PDL book, where we wrote down all we were responsible for to pass on to our successors. When I applied for an operations position, I was handed a copy of the current pilot’s PDL and also those of the operations assistant and the head of engineering. The current applicants were grilled so hard we practically had to memorize everything. Then, about a year before we expected to reach Alpha-1, we started daily practice sessions. This involved taking turns in the various positions while the other two gave us problem situations and then critiqued our performance.”

Larry posed another question, “What was the routine you used to orbit and land?

Norm answered quickly, "Oh, that's easy. We used exactly the steps recommended in the NSEA operations book. We trained ourselves, and then followed the instructions."

Larry answered, "And now you will see there have been a lot of changes, and these changes made things a lot easier. For our NSEA landing, I simply push buttons: The Landing button is first, then NSEA, and then Pad-3. You see, everything has already been programmed in the computer. Select the Orbit button and the computer puts us in orbit, largely by asking questions for altitude, ship attitude for vertical; and asks a question about the landing time. The computer places us directly over our landing pad and reduces our vertical speed so we have a soft landing at the center of our selected pad."

Norm commented, "Wow. That sounds like the computer makes everything easy."

Larry answered, "That's right. Now here's the catch. We have to perform a manual approach and landing as part of our tests to qualify for a pilot position. It has been five years since I did a manual approach to NSEA. Norm, my memory isn't that good. If our computer fails, I will need your help."

Norm replied, "I will need to look at your operations book to see if anything has been changed."

Larry answered, "The book is in my office. But you will have to do your reading there because the rules say it cannot be removed.

"Well, that's fine with me. This way we can spend some time together. Maybe that will be enough for me to help you memorize all this."

The next day, Larry asked Norm, "Would it be OK for my assistant, Jim Womack, to join for these training sessions."

Norm answered quickly, "Absolutely. But he needs to understand, we cannot approve him for a captain's position. That needs to be done by NSEA's flight instructors,

The next day was difficult because Jim did not have a full understanding of how things work. Larry started to reprimand Jim, then reversed himself, "Jim, it's not your fault. You have performed several lift-offs and many course changes, but you have never landed the ship. Your last flight test was really an approval to continue your training, which is where I should have helped. I'm sorry I didn't think of that earlier."

Jim agreed to spend the remaining hours today and tonight reading the particular sections of the operations manual and Norm would start his new training for Larry and Jim the following day.

The next morning, school was under way. Norm asked, “Jim, do you know the purpose of flipping the ship?” Jim replied, “Yes.”

Norm replied with “OK. But tell me why we flip the ship.”

Jim answered, “When we need to slow down, we first must turn the ship around, so the thrusters are in front.”

Norm asked the next question, ”Very good. Now, can you tell me how to do the flip?”

Jim took a deep breath, “Yes. I start by tapping the FLIP button. Then the computer asks if clockwise is OK, I tap YES. Then the computer asks if a five-minute turn is OK, and I tap YES. Then, the computer asks questions about the amount of thrust I want to use. If I am in a conservative mood, I could pick five or ten percent thrust and start the thrust one minute after the turn is complete. The computer then creates a chart showing me the percent relative speed we will have achieved at five-minute intervals.”

Norm answered, “That was great. You did a fine job with your description. You have done this before?”

Jim said, “Well yes, I have done flips on five trips. Three of those were with Larry.”

Norm decided we were done for this morning, “Jim worked hard and gave us an excellent summary of the flip process when the computer is working. Let's break for lunch and then we will continue."

For the afternoon school session, Norm said he wanted Larry and Jim to walk through the approach, orbit, and landing, all assuming the computer is available. Larry did most of the talking for these, and it was apparent he knew what he was doing. When they were finished, Norm complimented them both for doing such a good job in their presentations.

The next day. In the morning, Norm was prepared to discuss the four phases assuming the computer is not available. He assumed his presentations would probably take a full day, then they could start their "practice" sessions for a few days.

"Let me emphasize, whether the computer is available or not, the logic is the same. The real difference is knowing what is next and, sometimes, being ready to do some of the math, if necessary. It is wonderful that someone was able to do the programming so the computer can do the math for us. Again, we know what has to be done. You have been able to use the computer for the math. But now, if the computer fails us, we will have to make some decisions and do some math."

Norm has a special talent for flying a space cruiser, and he knows all the steps in each of the maneuvers. He is also quick with creating math formulas that need to be used. Where does he keep all this information? In his head, of course. And, for backup, he always has his PDL logbook.

After lunch, Norm gave Larry and Jim copies of his PDL logbook and said, "You are certainly free to use this to get started with your own PDL, but believe me, the notes you add to your book will eventually be your most important items."

Norm started with the flip. He carefully walked from step to step, using his own PDL and showing how each step matches with the process used with the computer. "Our goal is for you to store the details of each step into your brain. And it is very helpful to outline those steps in your PDL, so you can turn to that if you have a brain cramp."

"The flip is necessary when it's time to start slowing down. Our main thrusters are for accelerating and decelerating the ship. Yes, that includes both increasing our speed and decreasing. After we launch for a long trip, the ship will be pointed forward with the main thrusters blowing out the rear. If we want to coast along at some speed – we usually call that our cruise speed - we shut down the thrusters and the ship will continue at the same speed."

We are coasting. Inside the ship, everything is at zero G, and everything floats. Some think this is the fun part. Others hate it.

In closer trips we might spend three or four days accelerating, then cruise along for a while, and then later we will flip the ship and spend the same number of days slowing down or decelerating."

The remainder of our day consisted of walking Jim through the important events and explaining how we perform these events without the computer.

Norm spent two more days talking about the procedures preparing to land the ship, and then the details for the landing.

Wednesday evening, Larry spoke over the pri-comm to give everyone a brief on our schedule for tomorrow morning, "About ten in the morning, we should be starting our orbit around Planet Earth. We plan to be at an altitude of twenty miles, so if you remember your geography, you can figure out where we are. We will be close to our destination at twelve-thirty, and then start our descent about one o'clock. You will need to be in your seat belt several times. We will try to give you a five-minute warning each time, so pay attention. After we make the landing, it will take us at least half an hour to cool the engines and fix the ladders for us to exit."

"The ship's officers and special guests will be first to exit. There are about ten of us. The first half of the ship's crew, about forty people, will be next. The others have chores to do and should all be gone by about three hours after landing. Then we are done.

We shut down the engines and prepare to leave the ship. That's all I have for now. If you have any questions, you can send me a message on the AVC."

Larry and John headed to the cockpit for the night. Larry is the ship's captain, so he spends nights in the cockpit as the primary pilot. He uses John, his lead geologist, as back-up when he has an all-nighter. Larry, typically, started out on his duty for a couple of two-hour turns, and then shorter turns afterwards. John's duty was to serve two one-hour turns, and then, whatever Larry needs. When both were awake during quiet times, they usually played cribbage. Our landing is scheduled for two PM.

Thursday Morning. Their meeting time was scheduled for 8:00 AM, so John had arranged for six rounds of breakfast at 7:30. He also scheduled a large coffee pot at 8:00 with a large assortment of donuts and sweet rolls at 8:15, and then repeat that an hour later.

At 8:00, Jim walked over to Larry and said, "Good morning, Captain, how is it going?"

Larry answered, showing a touch of formality, "We are presently 40,200 miles from Planet Earth. Our present speed is 29,600 MPH, and at this speed we will reach Planet Earth a few minutes before 9:30. All gauges and other indicators are good." Larry stood.

Jim saluted and spoke, "I relieve you, sir."

Larry returned the salute and replied, "I stand relieved," and then with less formality, pointed to the chair, "Have a seat, Jim."

Jim took the captain's seat, smiled to all, and said, "My first action is to slow down the ship so we can arrive at Planet Earth at 10:00. I did the math and figured our new speed should be 20,100 MPH. If there are no objections, I am making that change now. Step one is to flip the ship." Jim changed to his formal voice, "Main thrusters off …now!" About a minute later, in his formal voice again, "Prepare to actuate the side thrusters forward on the port side and aft on the starboard.… execute now!"

Then in his casual voice, "I will be watching the heading for a 180-degree turn. Now we can turn off the side thrusters, … stand by … execute now!"

Back to his quiet voice, "Now we can prepare to reverse the side thrusters. Prepare to actuate forward starboard and aft port … in about three or four minutes."

Several minutes later the needle was at 175 degrees and moving slowly toward 180 degrees. "Stand by to turn on the designated side thrusters … stand by … get ready … execute. And in one minute we will turn off the side thrusters." A minute later, "Now stand by to turn off the side thrusters … stand by …execute."

Jim returned to his casual voice, "OK, the ship is flipped, so now we can slow down to 20,100 MPH, then we will flip back to normal and proceed to Planet Earth." While speaking, he turned on the main thrusters and moved the throttle up to eighty percent. About five minutes later he turned the thrusters off, and said, "And now we're at our cruising speed, and we can flip back to normal and continue this part of the trip."

"We are aiming for position at an altitude of forty miles, with a heading due east, and an increased speed of 26,000 MPH. That will put us into our orbit around Planet Earth. I would like for several of you to look at the surface and see if you can figure out where we are. I spread out the large navigation charts earlier. With those, see if you can give me a heading to take us to NSEA. It is located on the northeast coast of Florida."

During our first few minutes of orbit all we could see were clouds and water. We were over the Atlantic Ocean and heading east. Finally, the clouds cleared, and Larry hollered, "Look up ahead. I think that's the Strait of Gibraltar. We will be over the Mediterranean in a few minutes. On the left side – I should say on the north side – are Portugal and Spain. And then we will see France and Italy in a few minutes. On the south side we have Morocco, Algeria, Tunisia, and Libya."

We had a good time guessing the names and locations of all the countries. The navigational charts were a good help. But we really did not have much time to discuss anything. At 26,000 MPH we traveled the length of the Mediterranean in about five minutes. Another thirteen minutes and we have passed over China, Korea, and Japan, and now watching the Pacific Ocean, which must be five thousand miles across.

Larry said, "I've got it. When we cross over the western edge of California point the ship to a heading of 095 degrees."

The coast-to-coast trip over the US is about 3,000 miles, and our crossing took seven minutes. Jim called out to the others, "We are going to be fifteen minutes early. Apparently, we didn't spend enough time over the Atlantic Ocean. I was expecting to be over NSEA at 1:00, but we will be there at 12:45. I propose to continue heading east over the ocean for 7 minutes, then do a tight turn and fly another 7 or 8 minutes back to NSEA. Any objections to that?"

One response was, "Nope, that sounds good."

Another response, "Sorry we didn't catch that earlier. Good job on your part."

Jim took over again, "Here's what we're going to do next. About one or two minutes prior to passing over NSEA, I will start pulling our nose up. That will be a tight maneuver - about two Gs, and about one minute. Larry, can you inform everyone. "

"This is Larry, warning you that we will be making the turn upward in about five. This will be a 2-G maneuver, lasting about a minute. It may peak at 3-Gs, but not more than that. Put away everything that is loose and get yourself strapped down. The time is 12:45 and we are due to land at 2:00. And we are 15 miles above our landing pad.

Four and a half minutes later, "This is Larry again, and this is the last warning." Larry was quiet for a few seconds, then he saw Jim repositioning himself with his left hand on the throttle and his right on the control stick. "Three, two, one, here we go." Then Jim pulled back on the stick, and quickly adjusted the throttle to forty percent.

All was going well. Larry had a smile on his face, "And now, we just sit quietly. The maneuver will be over in two minutes. Then we maintain this position. The ship is pointed upward, with everyone laying on their backs. And now we are slowing the bird so we can have a nice landing."

Larry is now in the role of #2 pilot and takes his position where he is looking through the scope. When the front end is straight up, the "x" in the scope should be centered on the landing pod.

Jim announced with a happy voice, "Yay, the nose is pointed straight up. Our speed is still 66 MPH upward. We will be here for a couple more minutes, until I can figure out what throttle setting will hold us stationary. Hmmm, we are at 42 MPH already. We are slowing down. Now we're doing 32. I am adding a little throttle, so we don't slow down too soon. If we lose control of this, we could fall the rest of the way down. There. We are now 16 MPH and, up to 41% throttle. 8 MPH and still 41%. 4 MPH, and 42%. 1 MPH. There, zero MPH. 43%." One minute later, no change. That seems to be holding. Jim was sweating and asked for a towel.

"OK, we're starting down. Power is 42%,

Norm said, "I felt some movement. We moved."

Jim said, "Look there, we are headed down, stern first, and doing 2 MPH. We did our approach at an altitude of 16 miles. At this rate, we will take another 8 hours to get there. I'm adjusting the throttle back to 40%. Watch for our speed to increase. Larry, how are you doing?"

Larry said, "I see we are at the left edge of the pad, and a little outside the edge at the top. We are relying on aerodynamics, and they are useless without some air flow. We need some speed."

Jim said, "Got it. Throttle is down to 38%. We are speeding up. 7 mph. Now we have 10 mph. How is that. Larry."

Larry answered, “Oh, neat. I made the correction for the left side of our landing pad. Working on the front section now. That’s better. … Now more better. There, we are at dead center on the pad, and now I can keep it there.”

“Jim was thinking out loud. “Our altitude is 12 miles. The thrust is 55%, and the speed is 4 MPH. Look at the speed indicator. The slowest speed shown on the gauge is one foot per minute. That’s where we want to be when we touch down. OK, let’s get closer.

Two minutes later, “We are 10 miles up, speed at 10 MPH. We will need more than an hour and a half. Not good. We could go15 MPH. “Hey gang, we’re increasing to 15 MPH. Ten minutes later, we are closer, and have slowed to 5 feet per minute.“

Larry said to the crew; “**This is your last warning. … It’s time to brace, so hang on. We’re there. We have landed!** Larry, you’re now in charge” Jim grabbed a towel and mopped his face and neck.

Larry went to the mike to advise everyone that we are under lockdown for half an hour while fumes and hot air settle. “Our 3rd floor should be vacant except for those who are preparing to leave the ship. Right now, that includes our six people who represent Alpha-1, and the two NSEA employees. A half-hour later, we saw the truck stop at the edge of our pad. A few more minutes they moved closer and prepared to position a long ladder. Finally, our two large doors are open, and the top of the ladder is joined there. In another few minutes we are prepared for our exit.

An attractive young lady appeared and introduced herself as Diane, an assistant to the President of the National Space Exploration Association (NSEA). She then said, "The president is anxious to meet you, and he expects to be spending his time with you for the remainder of the afternoon. You can follow me, please."

It was a short walk to the president's car, a six-passenger limousine. Diane sat up front with the driver and the six of us were very comfortable in the rear seats. The president's residence was an attractive mansion, placed between the headquarters building and the landing pads. Diane said, "The president and I were watching from his top-side deck. By-the-way, he thought it was an impressive landing."

The car pulled up into the circular driveway, and Diane said, "You will be staying in the Visitor's Wing of the President's Mansion. You can wait here for a moment or two, and you will have an escort to take you upstairs. Other assistants will bring your baggage and any other personal items from your ship."

She started to walk off, then stopped, turned, and said," I will see you in the meeting room at three o'clock." And she was off.

The President's mansion is quite impressive. Our Sleeping Room was quite large. Actually. it is the entire wing of the third floor. Six nice beds and comfortable chairs. Dividers were provided here and there in case someone prefers privacy when changing clothes or sleeping. The shower room had ten shower heads so we could each control our own hot and cold adjustments. There were large mirrors and ten sinks and outlets for electric shavers. We would certainly have no complaints about these arrangements.

We had not brought any formal clothing, and Larry asked one of the assistants about our attire. The assistant said, "The president is currently outside playing tennis with his wife and will be coming to the meeting at three o'clock sharp. He is a casual dresser."

Someone asked," Then we will be OK the way we are? "

The assistant replied "He said your three o'clock meeting is casual and will last about an hour or two, no more than that. The purpose of this meeting will be for him to get to know you and share with you some of his thoughts."

"The President also asked me to pass on to you his plans for the evening meeting. That meeting will be in coat and tie and will include The President, his assistant, several of the senior members of the NSEA Board, and the six of you of course.

Later this evening he would like for everyone to gather in the dining room.at six-thirty. Dinner will be in the same room as your three o'clock meeting. The attire will be evening casual. That means coat and tie. Now look in this closet. There is a good collection of different styles and sizes. These are all left over from former passengers or crew members. We contacted these people and offered to mail them the clothes, but most were not interested. Typical responses, if any, were No thanks or give them to Goodwill. All of these have been cleaned and disinfected, so please rest easy on that subject."

We gathered in the meeting room at three o'clock and were made to feel comfortable. The president introduced himself and gave us a brief, but impressive, biography. He asked the six of us to do the same. All total, the bios took about thirty minutes.

Then The President shared with us some of the things that NSEA is considering. "First, you guys have been sitting on a gold mine. Yes, that includes gold; and it also includes many other valuable resources. Just as important, the resources at Alpha-1 are much easier to access. So, how do we mine the resources, store them in accessible places. and then load them in spaceships headed for Earth? You have plenty of people on Alpha-1, so I think we can post help wanted ads, and teach some of these how to be geologists. They could become employees of NSEA and receive a salary.

"According to some of the reports I have seen, we have thirty-five-hundred personnel currently living on Alpha-1. We have been playing with numbers for a while, and I am wondering if we could hire a thousand workers to help us scour the areas for minerals and other valuables."

Three hands were quickly raised, "Are the work areas very far from our homes?" "Can some portion of our salaries go to a retirement plan?" "How many hours a week?"

The President smiled, "Good questions. Regarding the first question, I think the areas would become pretty sparce over some number of years. So, let's say you might work within an hour's flight for the first ten years. Second, the answer is yes, you can have a retirement plan. You will also have a vacation plan and insurance. Third, I have been thinking in terms of forty-hour workweeks, with travel time considered as work hours. And we can work deals, such as four ten-hour workdays."

"Now you can answer my question. Do you think we can hire a thousand workers? But first, let me tell you why we are asking for this. Moving workers from one place to another is expensive. But you are already there. We may need to ask some of you to move to a new area, but that is not expensive. Throughout our areas we will be looking for metals, minerals, and gas."

A hand came up. "My name is Maynard Jones, and I am the Governor of Our Town and Western Village. Our head count is about 2700 and I believe we can provide half of our men so that should be about six or seven hundred:

Another hand, "I am Dave Parker, City Manager of the town two thousand miles to the east of Maynard's two towns. Our total population is eight hundred, and we can probably provide two hundred workers.

The President had a smile on his face, "So, we have an estimate of eight to nine hundred available to start, and that is great. We have another two or three days before your ship will head back to Alpha-1. That will be enough time for us to work up a plan to get ready for harvesting some of the valuable resources."

"We will provide ten technicians who are pad experts. I also want to get started with two buildings, so we will provide ten carpenters and other skills for the buildings. Counting the six of you, I believe that will be a total of twenty-six passengers on the ship. We will make some changes to the ship to carry a zoom-car, which is a small ship that can carry two pilots and twenty passengers. The zoom-car is made by the same company that developed the buzz-craft. The zoom-car will have a slower top speed, about eight hundred mph, but that is still fast enough."

The President looked at his watch and frowned, "Oh my, how time flies. It's almost five o'clock, and I have some things to take care of, and then get ready for our six-thirty dinner meeting. We will have some important people at the meeting. Remember, coats and ties, but nothing fancy. Gotta go. See you at six-thirty." A moment later, "I would like to have another casual meeting tomorrow meeting to work out some details,"

Back in our bedroom. We had a good time sprucing up in the shower and shaving in front of the mirrors. Then we had another good time selecting our shirts, jackets, and ties. Some of us had never seen the likes of these. Fortunately, two of the NSEA assistants were here to help us with our selections.

Now it is six-thirty and Diane has grouped us in a convenient area. "We will wait here for a few minutes." A server brought each of us a glass of white wine. Diane gave us some instructions, "This has alcohol in it, so just take a sip now and then. Too much of these and you'll start acting silly."

We noticed the NSEA officers were positioning in a line. Diane said we should each walk slowly down the line, shake everybody's hand, say some nice words, then step on to the next person. I will introduce you individually, one at a time, with names, space cruiser, and when you landed on Alpha-1."

We spent half an hour in the reception line, then we were seated. The President and his assistant, Diane. are at the head of the table, and we have three seats on each side.

Other officers present were: Chief Executive Officer (CEO), Financial Officer, Legal Officer, and. Flight Training. During dinner, conversations were mainly for getting acquainted.

After dinner, the President stood, took a deep breath, then said, "Welcome, welcome to all of you for attending this very special meeting. I have had several meetings with the senior members of our Board. We have a lot going on and some of these are beyond our experience. Many of the people from Alpha-1 have developed some good experience, but we still need to be prepared for new things, so we will need your help. We have several very important projects that have already been reviewed and approved. Let me start with a summary of those."

"First, we talked with the leaders from Alpha-1 about assembling teams for harvesting the valuable products. Their first estimate is six hundred to eight hundred workers, who are planning to start work within the first week of their return to Alpha-1."

"Second, we obtained their approval to install three landing pads near the two western towns, and two more pads for the eastern town. NSEA is providing a team of six to do the work for installing these pads. We estimate completion in about two weeks."

"Third, we also want to get started with two buildings. Each building will have meeting rooms and work rooms on the first floor, and apartments on the second floor for use by stay-overs. NSEA will provide four senior carpenters and we will invite another ten workers from the local towns. We will have sketches and cost estimates for final approval within the next two or three days. These plans have already been reviewed by the senior NSEA officers and these plans have been approved."

"We are providing two zoom-cars. These are nuclear-powered taxis that can carry up to twenty passengers. The zoom-car is made by the same company as the buzz-craft. We will make some changes to the spaceships to carry the zoom-cars."

Tuesday morning. Several of the senior personnel showed up for our breakfast meeting. We had a good time chatting with these important people. The President kicked off our meeting a few minutes after eight. "Our Legal Officer has already been working on this next one.'

The Legal Officer looks very good in his casual clothes. He started, "As you know, NSEA is an organization within the US Government. That means we cannot show a profit from our services or products. We can, however, temporarily provide a loan to a company for expenses as long as the loan is repaid within five years. We have looked into providing support to several new companies that can take over some of these functions. We have started to investigate a couple of small companies that could hire some of the Alpha-1 employees to harvest their products. Our CEO said we can drop that one. But if a small company is owned by Alpha-1 personnel, it works just fine. I suggest we consider helping some of the Alpha-1 personnel to create their new companies. This is easy to do. Now, you will need a president who knows about geology and a VP who can manage the workers.

The Legal Officer started to leave, and then he stopped. "I have some recommendations for you. Starting a small company is not that hard .You will need to send a letter to me describing what you want to do. Yes, I am the authority for what you want to do. For today, this is just for information. I will regard your letter as a draft. I will review your draft and then send you some helpful recommendations.

The President said, "Here is something to think about regarding my favorite subject. As some of us start talking about exploring other planets, I start thinking we could help those associated with our NSEA facilities on Alpha-1. That would make a nice jump-off to other planets in that direction. But . . . no need to hurry. That won't happen for a long while."

There was a knock on the door and one of the President's assistants entered and handed him an envelope. The President held his hand up to stop some of the chatter in the room. "Sorry for that, but I just received a message, and I am called to a meeting with the CEO. I'm sorry, this is for board members only, but I will try to give you an update when I come back. That will be a brief meeting. Can we continue with the rest of our meeting about three PM? Everyone said okay, and the President headed for his assistant's car sitting outside.

Two hours later, our group was ready to continue our last meeting. The President returned right at three, and he spoke, "You may have heard this at the six-thirty meeting yesterday. Someone said they are looking forward to watching us. The emphasis seemed to be referring to the day after arrival in Alpha-1. I quickly said we had already promised everyone the day off. Heads nodded okay, and we continued."

"Our second day should be primarily to confirm the leaders we have been thinking about and then starting to identify other good possibilities."

"During the remainder of the first week, we will be bringing in the new workers, signing them up, and taking them out for some first-class training. Starting on the first days of the second week we should be hard at work."

We should probably start two teams for the Western group, and one for Eastern. For our initial trial period, we will start with twenty people in each team. We will also need to have carts for carrying our collected metals and minerals, Also, we need a place to store our harvests for perhaps a month. The President said after our harvesters are hard at work, he might contact one or two outside companies and discuss possible contracts to periodically pack up the harvest and fly it back to earth.

Our target is to eventually have eight teams, and perhaps a hundred workers in each.

The President said, "Let's talk about pay scales. I'm not too concerned about the actual hourly rates, but we do need to have something to start with. You gave us a brief presentation of what you have, and I understand that is what you plan to use. Who wants to start your presentations?

Both Maynard and Norm raised their hands. The President pointed to the first hand. Maynard was delighted to have the President's attention, "Most of the residents of Our Town have been here for more than four hundred years. We settled in okay, quickly re-learned the skills we needed, and developed a school for our new youngsters. Higher education was not important to us, and our attention only went up to the sixth-grade level. And now, just ten years ago, Norm landed his ship five miles away from Our Town. Norm's people had a lot of time on their hands, so they required classes through the twelfth grade, and provided college-level courses to those who wanted to go further. These new people had a higher IQ, and they took over the combined education system and also helped us develop a finance system and a pay scale."

Maynard didn't need his notes, "Unskilled workers would receive hourly rates ranging from $5 to $10. Skilled workers' salaries might range from $10 to $20, with supervisors up to $25. Managers of larger organizations might receive salaries up to $35."

The President had a big smile, "This might be exactly what we need. How long has your pay scale been in operation?"

"About nine years."

"Great. And how many workers are covered under your pay scale?"

"About eight hundred workers."

The President's smile was growing larger, "This is fantastic. How about the rest of you. Are you satisfied with this?"

Norm said, "Maynard is great with this kind of thing. I certainly agree.

The others gave a head-nod or a thumbs-up. The President said, "I believe you have this under control. Let me know if you need any additional help."

Later that afternoon. Larry's ship will be carrying a much larger load than when they came to visit NSEA just three days earlier. When Larry checked this morning, he was afraid his return trip might have to wait another day. They flew here earlier with a light load, and today it will be very heavy.

The additional load includes twenty technicians to work on the five landing pads and the two buildings. Four NSEA mechanics have been busy disassembling the two zoom-cars and positioning them carefully on separate decks of the NSEA space-cruiser. Three other decks were updated to handle the twenty-two technicians to work on the five landing pads and the two buildings.

Wednesday. Final preparations for lift-off. Larry did a walk-through to satisfy himself that all was complete. Then he carefully reviewed his earlier allocation space for everything.

Floor 1 is the welcome room, with large double doors for receiving large materials. Elevators start there and stop at the Operations Section, with a level for meetings and also the Cockpit.

Floors 2, 3 and 4 are for the two zoom-cars, and some tools and spare parts.

Floors 5 and 6 are work areas for the ship's engineers and a good variety of tools.

Floor 7 is for those important Alpha-1people who made the trip.

Floors 8, 9, 10 are available as meeting rooms.

Floors 11 and 12 are for the recent graduates with degrees in geology.

Floors 13 and 14are for the NSEA carpenters for the buildings and landing pads.

Floors 15 and 16 are for the Pilot, Second Pt, Pt assistant, and cockpit.

All personnel for the above were advised of the various level assignments and to bring all of their luggage and any other heavy material. All shall be brought to the ship any time after five PM, but not later than. eight PM Wednesday evening. All concerned were welcome to stay overnight but will not be allowed to leave the ship. The final call for all others is to check in and settle on the ship by eight AM Thursday morning.

Chapter 10 – Back to Alpha-1

Day 1, Thursday Morning, Leaving Planet Earth. All personnel are here, and our internal inspection looked good. Norm is in the pilot's seat. and Larry has the mike,

Larry is speaking to Air Control, "NSEA control, this is SC253, checklists completed, and we request permission for lift-off."

"Roger, SC253, your sky is open, and you are cleared for lift-off."

And that was the end of our communications with anyone on Mother Earth. At least for a few days. Norm is the duty Pilot for this shift. Larry will have the twelve to four, and Jim the four to eight. That rotation will continue for the next twenty days.

The acceleration process was scheduled to start about an hour after lift-off and is being handled by Larry. A good part of the hour would be used for giving instruction, so Larry took a deep breath and turned on the mike.

Detailed instructions were sent to everyone this morning and appeared on all the AVCs. We used this for the first time on our trip to see NSEA, and it works great. But it requires all personnel to mind the safety signals.

Larry said, “OK, I see everyone is on-line. We will need you to stay tuned in. About fifteen minutes from now I will start the one-minute count-down. It is most important to make sure your ready light is on. If we see a ready light not on, we have to cancel, check out the problem, and start over. If we tried to execute the jump anyway, the force could slam that person to the back wall, and remain pasted there for the entire five minutes.”

Fifteen minutes later, Larry said to those in Operations, “I just handed each of you these off buttons. If you see anybody not participating or we have an emergency of any sort, we must turn the jump off. If someone hits off by mistake, do not attempt to cancel that by hitting the off button again. That can mess things up. We will have to wait for things to settle down, and then we start over. Time for the one-minute count-down was getting close. Larry turned to face Norm, and said, “Thumbs up for you?”

Norm showed both thumbs up, “Everything is OK. Let’s do it!”

Larry turned on the mike and said, “This is Larry speaking, and this is the one-minute count-down. Make sure you have no loose items in your bunker.” Then, less than a minute later. “. . . Three! . . . Two! . . . Hang on! . . . Now!” And Larry pushed the button to execute Jump 1.

Eight minutes later, Larry was back on the speaker, "You can relax now. That was a good one and we are now doing eighteen times the speed of light. You can relax for about ten minutes, and then I'll give you the countdowns for the second and third Jumps."

Jumps 2 and 3 are all done. Larry gave the all-clear on the phone, and then this quick summary, "The whole process took a little less than an hour, and our speed is now thirty-two times faster than the speed of light. You can enjoy the steady cruise speed for most of the rest of the trip. Now, you will remember our cruise speed is constant, so you can play 3-D basketball.

And, the day before we arrive, we will have to repeat the 3 jumps routine. We will flip the ship, so we are facing backwards, then we do the three jumps, and then flip the ship again. We have not yet decided if that should be on day 19 or day 20. But I will let you know when we are ready to put that on our schedule."

The three pilots had agreed to a simple schedule for their watch assignments, with Norm first, then Larry from noon until four, Jim four to eight, back to Norm for eight to twelve, and then continuing the same rotation for the remainder of the trip.

Frequently, the off pilots used cribbage to relax the boredom. Jim Womack had been Larry's operations. officer for three or four years. Jim is one of those really smart people, and Larry had been using him as his pilot assistant for at least two years. During the flight for our visit with NSEA, Norm spent several days with some of his personal training so Jim could soon pass his pilot tests. He is doing just fine.

Larry may not be able to play much cribbage. He seems to always have something important to take care of and he works on those during his relief.

Day 2. Early the next morning Larry did a quick survey and he saw that this floor had enough tables and chairs, but it still has all the apparatus for basketball.

Larry called Melvin, "Hey Melvin, I gave you floor 10 for your meeting this morning. I just took a look, and it has 8 tables and 40 chairs. That should be all you need, right?

"That's just fine. But you called. Is there a problem?"

"Well, that floor also has everything it needs for basketball."

"Hey that's great! We can switch to basketball when our meetings are over. You know, I was the best on Norm's ship about ten or twenty years ago, when I was thirty to forty years old. And Norm was the best before me. I love it."

“I’m glad you’re so happy. But we might have to kick basket players out so you can have your meeting. How about we give you floor 9 for the graduates? then everyone can be happy.

Melvin replied, “I am sure you are correct, but I have another thought.”

“And what is that?”

”I would like to have our first meeting on floor 10. And then, after the meeting is finished, we can play some basketball. We can have some fun for a few minutes, and I can teach them 3-D basketball. And then we will use floor 9 for all of our meetings after this one. This should be a better way to make everyone happy.

Larry replied, “That’s a deal. I like doing business with you.”

The First Geology Meeting. Attending are the 22 graduates, Melvin, and Larry. Melvin went to the podium, “Hi Gang, I’m Melvin Sherwood and I have met some of you in earlier conversations. I have a lot to share with you this morning, but first I want you to hear from our Pilot and ship Captain, Larry Anderson.

Larry took his place, “Good morning, everyone, and welcome to the ranks of NSEA. In case you forget, that is short for the National Space Exploration Administration. I am a NSEA employee and have been a pilot for our space cruisers for 5 years. I am not a geologist. I just transport geologists around so they can do their job. Some of the natural supplies from Mother Earth are starting to dwindle and they are harder to find. Our mission is to start finding more of these resources so we can ship to NSEA. Now, I have some other matters to take care of, so I am turning the podium over to Melvin who is our Senior Geologist."

Melvin took over, "Thank you Larry." There was a short applause for Larry, who waved, then disappeared into the elevator. Then Melvin turned his head toward the Grads. "Before we get into questions and answers, I want to talk to you about why we are here. Larry told you our mission is to research and find the resources that we can have shipped to Mother Earth. NSEA has been working on this for more than 400 years."

"They launched their first ship, Space Cruiser SC12, for this mission with 225 personnel to a planet named Alpha-1 just over 410 years ago. They had carefully selected a group that had the capabilities to search the planet, looking for these resources. Before they launched, they had to swear they would perform the mission. The whole group stood, participated in the swear, and then signed the papers over this; "*So Help Me God.*"

“The ship had some unexpected problems. NSEA lost connection with this ship and eventually assumed it was lost forever.”

“Two years later, they launched SC15, with another 225 personnel, and with the same mission. This ship made the flight and the landing. Unfortunately, it was a hard landing, and they lost their communications. They quickly tried to send a message they needed help, but nothing was received by NSEA. NSEA tried sending a message to SC15, but no messages were received. NSEA set about putting together a team to send there on a search and rescue mission.

It took them 3 or 4 months to prepare SC16 and locate 50 people for this search and rescue mission. All went well except for the landing. Unfortunately, this was a new pilot, who passed out when preparing to land. The assistant pilot attempted to finish the landing, but mixed the left and right signals and the landing was on the side of a hill instead of a flat area. The ship stood there long enough for everyone to jump out just in time. A few minutes later, the ship tumbled down the hill. It was unflyable and communication was completely lost. The last message received by NSEA was, “We are landing,” After 3 months of listening, they assumed the landing was a crash and no one survived.”

“NSEA lost the personnel from three ships, and was nearly out of business from paying a fortune to the families of those who were now dead, and they completely discontinued their plans for Alpha-1,"

"Now, 400-plus years later, the dust settled and NSEA decided to give this another try. Larry is our pilot and John Levinger is a Junior Geologist. He is 7 years ahead of you recent graduates "

"By the way, I am a senior geologist, with 15 years' experience teaching geology.

Several hands went up. One of the guys said, "I have a question."

Melvin said, "Finally, we have a question. Go ahead."

The recent grad spoke, "I understand you were born on one of the space cruisers and remained there until a short time ago. Where did you have your schooling?"

Melvin replied, "We had a school on our ship. This all started with some professional teachers on the ship. Our school went the way through college courses. After high school, I initially thought I would go for a degree in math or physics, and I included some courses in other subjects, just to test the waters, so to speak. Well, I fell in love with geology, and I found I had a good knack for it. I ended up with a masters' degree in geology, with minors in math and physics. After college, I started teaching math and physics to high school students. After four or five years, they were looking for someone to teach a college course in geology. A few days later, I talked to the recruiter, and then several of the current professors. They asked me a million questions, and then made me take a series of current exams. I passed everything and then taught geology for 10 years.. And now, for the past ten years, I have been the senior geologist for our crew as well as those from SC15, who have been living here for four-hundred years."

Our Alpha-1 people are in two areas, West and East. Our West area has a total of 2700 citizens and up to 700 available to learn some geology. The East side is smaller, with 800 people, and 200 of those may be available. We are hoping to start putting together some teams of our untrained volunteers, mixed in with you educated guys. You are geology graduates, you understand what has to be done, and you can help train our beginners. You signed contracts for a two-month period. We are prepared to provide your return to Planet Earth, at no cost to you of course. But, if someone would like to stay for a longer period, we will certainly consider that.

Melvin gave a big smile, then he continued, "Well, that's enough of that. Do we have any more questions?

Another hand went up.

Melvin, "And what is your question?"

"What are the resources we will be trying to find?

"Well, it is a long list, but here are the most common: Metals, including aluminum, silver, copper, magnesium, and iron; Others are: coal, oil and gas; rocks with metals, and valuables, including diamonds and gold.

"OK, no questions, great. This floor used to be busy with basketball players. As you can see, this is the small version, with only one basket. In order to play regular basketball, the ship has to be flying at one G, which gives us a gravity, like we had on Planet Earth, and we are very comfortable. At those times, people can play regular basketball."

"But we are not flying at one G. We will be at a very fast cruise speed all of the time. We are at a constant speed now, normally called cruise speed, and things float. The ball floats, and we float. You can see it is easy to toss the ball and hit the ceiling. We are in a three-dimensional condition, and we consider the ceiling is out of bounds. We have some other rules, and we call this 3-D basketball.

So how do we play basketball when the ball floats? You will want to try some of these different tactics. You will learn to think of it this way. Say you start at the far end, with the ball in your hands. You can throw the ball toward the basket, leap to the same place, grab the ball and put it in the basket."

"Or this. Back to the far end, dribble to the other end, turn around and then, with basketball in hand, leap for the basket and try to score. Or if a teammate is near the basket, throw to him and have him leap and put the ball in the basket. Some things are complicated. Holding on to the basket, or the backboard, is a penalty. Hanging on to the ceiling is a penalty."

Melvin had each of the grads try one or two of the maneuvers. Two of them made baskets. They both received good applause. We were having a good time.

Melvin blew his whistle. "Hey guys, we are out of time, and we have to leave here. Our remaining meetings will have to be on floor 9. When those meetings are over you can take a look to see if floor 10 is available. You are free to go now, and I will look forward to our next meeting on Day 4.

Day 3, Meeting for the NSEA Carpenters. The NSEA carpenters arrived for their meeting at nine this morning. This is Larry's presentation, and he is standing behind the podium. He waved his hands for quiet, and said, "Take your seats please." Short pause. "Welcome, gentlemen. I am happy to have this opportunity to talk to you and answer any questions you may have."

"Now understand, I am not a carpenter, and the closest I came was to make a doghouse for the family pet when I was twelve years old. My folks thought maybe I would go into woodworking, but that didn't work out. Instead, I went into engineering and later learned to fly these space cruisers. Now I get to fly you guys to your jobs from time to time. "

"For this morning, we will be reviewing what you will be doing for us in the next couple of weeks. But before that, I want to introduce some important people you will come to know. I was going to introduce Norman Link, but he is currently in our ship's pilot seat, and I will relieve him at noon today for the next four-hour shift. Norm was pilot of the SC12, which he landed on Alpha-1 ten years ago. By the way, Norm is probably the best pilot ever, as well as the brightest man in the whole group. "

"The next important person is Melvin Sherwood, our senior Geologist, He is a very important man, and the leader of all those people searching for and picking up resources for sending to NSEA to share with Planet Earth. Before landing on Alpha-1 on Norm's team, he taught college-level geology for fifteen years."

"And me? I fly this ship. And after I arrived here, a month ago, I released our buzz-Craft, a small, four-seated scooter, with a six-seat side car. We have two areas. Our West and East sides are 2500 miles apart. The buzz-Craft has a top speed of 4,000 MPH. That was very useful."

"Each area, East and West, has recently built airfields. West has a helicopter and should have an airplane within a couple of months. Dave Parker, City Manager at East, told me his engineers are working on a remote-controlled airplane."

"Inside our ship are two new zoom-cars, larger than the buzz-Craft that carries up to twenty passengers. The top speed is 800 mph, less than half of the other, but 800 is just fine."

"Now, the most time-critical assignment is to have a landing pad at both East and West locations. Landing these space-cruisers in the dirt is not always a successful landing."

"East and West will eventually have more. But that comes later. Also, you are authorized to start the two NSEA buildings. We have people searching now for who can start supplying you with building supplies, and we might have that information before we land. I understand you already have copies of the approved blueprints. "

"Well, time is up, and I don't want to be late relieving Norm. If you have questions, send me a note in your ACV. I will reply, hopefully with your answers. Goodbye for now."

Day 19, The 20-day period was wonderful. We have 22 recent graduates, 11 carpenters and a dozen or so of the rest of us. The grads and carpenters kept running into each other. They would usually stop and talk for a while, "*What do you do*?" "*Gee, that's interesting.*" Most of these carpenters are professionals who build houses, and spend their time in Florida, near NSEA, and frequently accept temporary contracts."

Larry and Melvin were able to provide these new employees with a view of what they will be doing in this new place named Alpha-1.

Preparations for the first week. Larry focused on the first few days only and said he would work on the other days later. Most important for the first day is to get East and West started in contacting resources for supplies for the building. Second, to start preparing at least one landing pad for each, He picked a small group to be responsible for removing the two zoom-cars from his ship. These have a speed limit of 800 mph and carry 20 passengers, so they are just right for carrying geologists. Larry is the only one qualified to be the pilot, so he will have to train our two geology leaders, Melvin and Jim. Larry's buzz-craft carries 3 passengers, plus 6 more with sidecar attached, and the high speed goes up to 4000 mph. Norm is qualified.

In the meantime, Melvin and John have charge of the West group geologists. On the first day Larry will need to take them up in a zoom-car for lessons. Then the next day they can start taking the geologists to the field. Melvin and John can take 20 each in the zoom-cars.

Other principal contacts are Tommy Smith (Mayor of Our Town), and Tom Kennedy (Mayor of West Village).

The 22 grads had brought several large tents to use for temporary home-stake until an apartment building is finished in West Village. Completion of that building is expected in a week or two.

Day 1, Landing Day. Larry had manipulated speeds so the cruiser could be in orbit not later than eight in the morning. That way he could give everyone an overhead view of the East and West facilities. Last night, Larry offered the pilot seat to Norm. Surprisingly, He declined.

"Larry, thank you for the offer, but I feel I should decline. Thank you for sharing enough with me. I believe you should give this assignment to Jim. Remember he earned some compliments when he landed us at NSEA. This will be great for Jim, and it will put him in better shape to pass the NSEA official pilot tests,"

Larry replied, "Norm, I expected you would take my offer, but I guess I am not surprised. You are certainly a super gentleman. I will tell Jim he has the assignment."

Norm has the four-to-eight Pilot schedule.

At seven AM, Norm said, "Jim, we need to change places. If I stay here until eight, we will be changing the watch at the same time we are doing the flyover. That's just too busy and not very safe. Let's change now!" *Pause* "Here is the turnover info. We are currently at 600 mph and an altitude of 5000 feet. I recommend 300 mph and 1500 feet over the flyover, then back up to 5000 feet and speed at 4000 mph. Then do the flyover at East, and then circle back to West for the landing."

Jim replied, "Got it. Norm, I relieve you. And I thank you."

Norm and Jim quickly shook hands, then Norm stood and said, "I stand relieved."

Larry was ready, with the mike in his hands. He announced in the Pri-Comm, "We will be passing over West in a few minutes. We will be at a low altitude, so you can have a good view. Then we will head for East. That's just for our fly-over. Then we turn back to West for our landing.

Ten minutes later, Jim lowered speed and altitude as Norm suggested. Then he pushed the throttle and did the flyover for East and then back to West for the landing,

Larry had the mike again, "Well gang, you can start bracing up. We will be landing at West-Air in a few minutes. But you should not jump out until I tell you it is safe."

There was a slight breeze, so just a few minutes later, Larry hollered to his passengers “It is OK to come out now, but be careful. Larry needs to move the two zoom-cars from here to the hangar on the other side of the airport.”

Larry phoned David, the helicopter pilot, “Hi Dave, I have 2 zoom-cars to fly to the hangar. Is that okay?

David’s reply, “No. you must phone the Air Control lady for permission. She is easy to talk to but will have your butt if you go to the air without her permission.”

“Thanks for the warning.”

Larry phoned Air Control. “This is West-Air Control, I guess that’s you, Larry. We had several conversations when you and some others were working for approval to use part of West-Air.” Larry replied, “I remember you, about two months ago. You were actually a great help to us. Right now, I need to relocate one bird to the hangar, and then another one to the East. OK?”

She replied, “This is Air Control, and that is approved. Say, that landing of the giant bird was beautiful.” Larry replied, “Not me. That was Jim Womack, my pilot assistant, and who will be applying soon for his own pilot license.”

Larry identified the nine carpenters he would deliver to the East, and explained he had to take the first zoom-car to the hangar on the other side of the airport, then he would be back to take them to East.

David brought Larry back in his helo, as planned. Then Larry loaded the 9 carpenters and explained the rules. Then he called Air Control for an OK. All was good and away they went to East. It is 10:00.

West. Melvin had talked, yesterday, to Tom Kennedy, Mayor of West Village, for contacting some suppliers that could provide us supplies for building our new NSEA building, "And can you arrange a meeting place, preferably at your office? And tell them we have pictures and details. Those guys are sharp."

Later today, Tom picked up the three professional carpenters and drove them to his office. The carpenters showed their pictures and details and gave Tom a quick overview. Then Tom looked at ads for what they needed. Tom had discussions with just about all those in West Village and Our Town. Tom reviewed their ads, looking for mention of Supplies. No help. He called the number of a man he had talked with several times, not too long ago. After a brief chat, they got down to business. He learned that his buddy builds houses and other buildings, and he relies on other companies for supplies. When finished, Tom had the names, addresses, and phone numbers of two who are highly recommended supply companies.

Tom called the suppliers, explained what we wanted, and invited them for a meeting at his office. Both were interested and they quickly arrived at Tom's office. Tom welcomed them and turned the meeting over to the leader from NSEA. This went well. The two suppliers identified the items they carry and provided their prices for each. We then had a good list, but still short for several of the necessary items. One of the suppliers provided contact information for two more suppliers.

Tom made the call and these suppliers arrived in a few minutes. They used the same procedure and then were pleased to see they had all of the requirements covered. Done.

Tom and the two leaders prepared their plans. Then they decided to share this with Larry.

About 3:00, Tom phoned Larry to tell him what they accomplished. He asked Larry if he wanted to see this before they continue,

Larry replied, "No, it sounds like you did an excellent job. The senior carpenter is the final approval for this, and we have no reason to wait for me. Hey, I'm headed for East. They are talking to a local carpenter right now. This is a good day."

East. On the road, headed East, Larry called Dave Parker, the East City Manager. He was with us when we made the trip to Earth for the NSEA visit, so he has a good understanding of what we have to do. Larry called Dave, "I just left and will be there in a half-hour, and our nine NSEA carpenters are with me. Most important for today is to invite a Supplies person to your office so we can arrange to order our necessary supplies."

Dave replied, "You know, all my men are carpenters or engineers, so that makes my job easy. I have already called the first one, and he will be here before you. He advertises as, *Micky, the best in town*, so I started with him."

Larry, "Great. See you soon."

Just half an hour later, Dave called Larry, "Micky is here. He wants to be the leader of the team. I told him we already gave the lead to one of the NSEA professionals. I told him this position may be open in a week or so, and we may have an opportunity to assign a replacement. Micky will be at the top of my list. I also told him we need a meeting this afternoon. He said my office is OK with him and he will be here. In fact, he is on the way, and might be here before you. I am certainly looking forward to this."

Larry replied "Thank you! I will be landing there in about ten minutes. I'am in my buzz-Craft so I have to slow down to drive on your bumpy road, so that will be another fifteen. Tell him we have pictures and details he will want to look at. That was an excellent job. I'll see you soon."

A few minutes later, Larry Landed in East's area and drove up the cobblestone road to Dave's office. Larry introduced the nine professional carpenters he brought, and Dave introduced Micky. They brought out NSEA's pictures and details, showing very adequately what the finished building will look like.

Dave commented, "We have been talking about who can provide the supplies we will need."

Three candidate suppliers showed up and their discussions began. Two hours later, they signed up two of the suppliers. The third was not what they needed, and he was dismissed.

Dave leaned on Mickey to "dig further" and find someone who can do this. Pretty soon, they were ready to give up. Dave turned to Larry "This is a small town. Any ideas?"

Larry replied, "I was thinking maybe we could bring some items from West in my zoom-car. How heavy and how big are they?"

The carpenters gave a good description of the four different items.

Larry smiled, and said, "That's no problem. Now, how many trips will I have to make?"

The lead carpenter said, "These are usually not high-quality items, but sometimes we have needed another 10 or 15 pieces. I believe a single load with ten pieces each might cover this, and that is easily just one trip. We may need a few more, but that should be easy to handle.'

Larry replied, "I will be glad to make the trip. Dave, would you talk to Tom Kennedy about ordering the items, then notify me after receipt so I can deliver to East?

Day 2. The plan for today is to take two flights of geologists out for our first practice. Melvin and John are the leaders. But there is a problem. Melvin and John have never driven a zoom-car. So, for today, Larry and Norm will be the pilots, and Larry will give lessons to Melvin and John later this afternoon.

The zoom-car has five rows of seats behind the pilot. The inside is similar to a bus. The pilot is at the front, with the door to his right. The rest of the zoom-car has five rows, with four seats each, and a passage down the center. Comfortable room for 20 passengers

Everybody gathered at the airport location at nine this morning, Larry, Norm, Melvin, and John, the 22 graduates, and 14 of our volunteers who have participated in previous occasions looking for geologic samples.

Larry and Norm will be the pilots to deliver the others to the area that were picked by Melvin. This was set up as a 3-hour event, so Larry and Norm will be back to pick up the group at noon. Melvin is telling the others what we are doing. "Any geology teacher will tell you the worst place to look at is flat land, covered with grass. So, why are we here? That's because those birds can't land where I'm taking you. See those trees over there? Just a hundred yards away is a stream flowing down from those hills in the background. That is a good area to stay in for the next three hours. That stream is probably full of rocks, the kind that our engineers make iron and other metals out of. We gave everyone a carry bag. Certainly, it is good to put some samples in there. But don't put too many rocks in there. We don't want to handle a ton of rocks.

It is noon, and two zoom-cars just landed. No one was in the mood to spend time telling what they did. Some wanted to go home. Some were just hungry. Within five minutes they were back in the air, and in another fifteen minutes they were landing at West. Tom Kennedy, Tommy Smith and Maynard Jones ordered fifty meals and six picnic tables. No one had any complaints about that. They had some great discussions, many of which were directed toward doing things right, not wrong. Some were taking notes.

Larry quietly motioned to Melvin and John. Then away from the crowd, they agreed this is a good time to learn how to fly the zoom-car. In an earlier conversation, Norm said he wants to be the teacher for Jim. That was fine with Larry.

Larry, Melvin, and John climbed into a zoom-car. Larry pointed to the two seats in the first row, both on the right side. “That way, we are just a few feet apart, and easy for me to see you. And make sure you fasten your seat belts,”

Larry pointed out the important things. The zoom-car has two engines. When the pilot is ready, he pushes the START button, and the first engine starts. When he is ready to lift off, he pushes the BLADE button, The engine shifts to the five helicopter-type blades which open from the bottom of the ship, turning at a medium-slow speed. The pilot controls the blade speed with the throttle and now carefully increases the speed, and the zoom-car starts lifting.

“In *city* locations, like this, I like to go up to ten feet. That way we don’t cut any heads off. Then we push the JET button, and now it is the second throttle to control our speed. Just a slight push, and the speed starts building up. Now I can turn off the first engine. We have a steering wheel, just like a car. A little left turn here so we miss those trees and tall buildings. Our speed is now 45 mph. That’s too slow, so we increase the speed. Now, we hold the throttle. See that. In just a couple of minutes we are at 120 mph. Now pull back on the wheel and the nose of the ship starts moving upward. Now add some more throttle, pull back – carefully - on the wheel. Just a small amount. Two more minutes and we are at 280 mph and at 4000 feet."

After answering a few more questions, Larry said, "We are flying to the same area where we landed this morning for the geology event. We will land in a few minutes. I move the throttle back, move the wheel down, just a little, and we are at 1000 feet. Then back on the throttle. stay at about 100 mph. I will fly over the location, then do a circle, dropping to 100 feet, then 40 feet, and slow to 40 mph. We must be pointed to the spot where we will land, slowly moving the throttle back all the way, and we feel a soft bump. We landed."

Larry pointed to Melvin and motioned for him to take the pilot's seat. Melvin sat down, then buckled in. He asked Larry, "Am I cleared to go now?" Larry answered, "Yes you are cleared. Just take it slow."

Melvin pushed the start button and the first engine started. He pushed the blade button and they started. He pushed the throttle, and we started moving upward. He pushed the jet button and adjusted the throttle . Then he turned off the first engine, pulled back on the wheel and pushed on the jet throttle. We were climbing and increasing speed. We are at 4000 feet and 400 mph. Just a few minutes later, Melvin said, "I see our landing spot up ahead, so I'm starting the approach." He started maneuvering to 60 feet and slowing to 60 mph.

Melvin continued "lower and slower" in the rest of the circle, and then, pointing directly at the spot where Larry landed. We are now at 30 feet and 30 mph. They continued slowing and lowering, with the "bump " close beside Larry's marks. Melvin released his seat belt, stepped back to Larry and said, "I did it! That felt pretty good. What do you think, boss"?

Larry replied, "I think that was wonderful. You must have a good teacher."

They laughed. Then Larry turned to John, "OK John, now it's your turn. "I hope you can do better than he did."

John answered, I will do my best, but I have never beat Melvin at anything."

Fifteen minutes later, John completed his flight. Larry clapped and said, "John, that was great. I wonder if you guys are after my job. I think I'll fly us back to West." Then Larry took the pilot's seat and off they went. A few minutes later Larry said, we will be keeping the zoom-cars in the hangar, at the other side of the airport. It might be easier for you to drive to the hangar, leave your car there, and then drive your zoom-car to pick up your geology teams at the other side. Our geology graduates will be living there in a large tent and may move to an apartment in West Village later.

A few miles from the airport, Larry phoned Air Control, "Hello, this is Larry in the zoom-car, I need put this in the hangar, then I'll go to the other side in my buzz-craft."

He landed and drove to the front of the hangar. Then a worker motioned where he wanted Larry to park. Larry and his two students jumped out. Larry led them to his buzz-craft and said, "OK, have a seat in the back, buckle up, and we cross over the airfield. Then we're done for today".

Most of the recent graduates were there and gave them a nice welcome.

Melvin stepped forward to give everyone a brief summary of tomorrow's schedule. "

First, the two zoom-cars will be parked in the hangar, over there. John and I will drive them over so you can jump aboard at the appropriate time. If any of you live closer to the hangar than this location, you might want to talk to your pilot and see if that works with them,

Day 3. Melvin gathered his gang and gave them a brief on today's schedule., "Good morning, guys. Welcome to our first full-day work shift. We gather here at eight and that is the start of this shift. We will fly to the same area as yesterday. Yesterday, we gave everyone a carry bag for carrying your findings. I checked around to procure forty back-packs. Nobody had any in stock. I placed an order, and we should have those next week.

In a few minutes, you will find a lunch package on your chairs. We asked for a variety on the sandwiches. One third of the sandwiches has red meat, another third has chicken, and the other sandwiches have fish. If you do not like what is in your package check around with your neighbors for a swap.

We recommend 15-minute mid-morning and mid-afternoon rest periods. Then the lunch period will be half-an-hour starting at noon. We plan for our two birds to head back west at 4:00. So, try to be back at your bird by 4:00. In the event you think you will be late, phone your pilot so he doesn't leave without you.

Day 4. The geologists carefully followed the plan. During the lunch period, the members talked about how much they are learning. Four hours later, and in the zoom-cars, they mumbled they were tired and worn out.

After landing, Melvin asked everyone to join him for a 15-minute wrap,

Melvin started, "I want to thank everyone for following all the plans. Today is Friday. Saturday and Sunday are off days. Then, starting Monday morning at 8:00, we assemble here for the same schedule. We will start advertising to the public in a few days. Then in the week after next, I expect we will be responding to our customer's desires."

In the latter part of the week, we provided four carpenters and several flights with loads of a variety of boards and other miscellaneous stuff. They were putting up with what looked like a farmer's barn. About 3 o'clock Friday afternoon, Melvin had called all of them to the new 'barn' the carpenters had built. They all took a peek inside, and wondered what animals would be there when they came back on Monday.

He circled the group and told them, "No, it is not a farmer's barn. This is where you can deliver the results of your work. We earlier provided everyone with carry bags. On Monday we will also have backpacks for those who ask for them. We will also have several wheel-carts to help you carry the heavy loads. With all these changes we will be looking like professionals."

Melvin continued, "Starting Monday, we will have a geology expert in the barn, who will tell you where to put what. I believe the carpenter will be finished with his tasks in a day or two and I hope the geology expert can be dismissed by the end of the week. He is expensive.

Later, "There are a number of locations on Planet Earth where geologists found very large chunks of iron. I do remember a location about a hundred miles west of the East location. I went down into the remains of a dried-up volcano. I didn't find any huge chunks of iron, but I did bring back a number of small chunks, each about the size of a baseball."

Day 5, East. Larry flew over here in his buzz-craft this morning and landed just outside the 'temporary' hangar. The foundation of the NSEA building is about twenty yards away, with the two landing pads another twenty yards.

The 'hangar' had been planned as an addition to the factory that has been there, successfully, for many years, and specifically for the purpose of building airplanes. In exchange they would be adding the new part behind the main part of the factory and the hangar. Dave Wilkin made this later suggestion and his lead engineers and factory manager agreed.

This is a Saturday, and the NSEA leader said their carpenters will be working six-day workweeks until these buildings are complete. The foundation appears to be finished and markings are being made for the inner posts to support the heavy beams that will come next.

Larry thanked everyone, then said, "Call me if you need anything." Then he started his buzz-craft and headed West.

West. With his destination in sight, Larry phoned Air Control "Hello, this is Larry, coming in from East, I need to land by the new NSEA building. OK?"

"Hi Larry, you are cleared."

Larry did a low circle, then landed, parked, and walked a few yards closer to the action.

Tom Kennedy asked, "How is Dave, and how is his building?"

Larry answered, "Dave wanted me to pass his good wishes to you. And his building is a little further along. He has the foundation done and they are preparing to put in the inner posts. But remember they have more NSEA carpenters than you do,

Back in NSEA HQ. The NSEA President, otherwise known in some circles as Harold Phillips has been a very successful businessman. Several years ago, when Harold was still new with NSEA, and a beginning geologist, he developed a relationship with Jerry, the local magazine's manager for articles. There was a "special" section for articles about geology. Some geologists might advertise, "We currently have an overload of *such-and such*, and we're offering these at a lower price. If interested, give me a call."

Occasionally they had articles placed by engineers, or engineering factories looking for *such and such*. Harold had seen several ads for excessive quantities. NSEA is part of the government and he had to be careful not to take a profit. Harold discussed this with the Board's people who understood all the details of these rules and they helped him understand what he can and cannot include in the cost. In his ads he planned to include statements like, "These prices are cost only, No profit."

Several years ago, when Harold was still new with NSEA; and just a beginning geologist, he developed a relationship with Jerry, the magazine's manager for farm-related articles. They continued to chat once or twice a year. Harold phoned Jerry's number.

Jerry answered, "Hi Harold, We haven't talked for a year or so. How are things on Alpha-1. And how did they treat you on Planet Earth?"

Harold replied, "Well, I think we had a wonderful visit with the seniors at NSEA. All things considered; we might eventually be able to earn a profit someday. That is, if we can get things started just right. "

"And what do you mean with that?"

Harold replied, "Well, consider this. We used to think that the resources on Planet Earth are about as good as they can be. But the full-time geologists have demonstrated slight decreases of the important sources. Well, the natural resources on Alpha-1 are at least double those on Earth. Our biggest problem will be shuffling people back and forth. Do you remember the first trips to Alpha-1? Back then a trip to Alpha-1 took nearly a year, each way, and today the trip is only twenty days. I talked to NSEA about helping. They are providing one space cruiser, including a pilot and two assistant pilots; at no charge, for the first year."

"Eventually, Jerry offered some suggestions for answering requests, "You need to be able to give an immediate answer about what you can do and what you cannot. Those items that you know are available are considered *yes*. Others are *no* (for now). Then, Later, you may want to search for the *no's* in new areas. Find them and they become *new yes's*."

Proposal to the NSEA Board. At the next NSEA Board meeting, Howard submitted to the Board members that our President wishes to be temporarily *retired* from his President position, and transferred to Alpha-1, perhaps for the entire first year. The Board members seemed to not know how they should vote.

CEO took the floor, "I'm sure not all of you know of the problems we had starting some 400 years ago. We were trying to get something started on Alpha-1, hopefully to find resources that might be of value to us on Planet Earth.

SC12. In the year 2065 we launched SC12, with 225 smart personnel, plus 25 ship personnel. Within a week we found out our SC12 was full of problems. Eventually, we lost communication and we eventually assumed they were lost in space. To everyone's surprise, they were able to live through a 400-year trip to Alpha-1. More on this later

SC15. Of course, we started planning for another try. We gave them the same mission as for SC12, picked another team of 225 smart people, inspected the cruise ship, and SC15 launched in 2067, two years after SC12. We later received messages describing their arrival, orbiting, and preparing to land. All went well and all personnel spent some time checking out the scenery. Communication ended and they blamed a bumpy landing. They thought about flying back but that was vetoed because they may need some repairs due to the 'bumpy' landing. The personnel eventually settled in and started building a new life.

SC16. "NSEA then moved quickly to prepare another team and another ship, this one with a double pre-flight inspection. The mission of this flight was search and rescue. They had a total of 50 personnel. The only pilot available had only recently completed his flight training. He was signed off as qualified, but we learned later he had some problems. During the landing phase at Alpha-1, the new pilot was so nervous he fainted. Another staff member said he could make the landing and took the captain's seat. Unfortunately, he had the left and right buttons reversed and ended up landing the ship on a hillside. The ship stood there for an hour or two and then tumbled over and rolled the rest of the way down the hill. Fortunately, they had enough time to abandon ship, and no one was hurt. They were good people and did not waste any time waiting for a rescue.

Back to SC12. "Well, this ship was not drifting in space. The operations computer had been programmed to fly to Alpha-1 and then release control so the pilot could circle around the planet and look for a place to land. When the original staff realized they would be on the ship for 400 years, or 20 generations, the many discussions to create plans and procedures were helpful."

"They had been very serious about carrying out the original mission. After all, they had signed a piece of paper by which they committed themselves to do their best to carry out the mission, *SO HELP ME GOD*. They created a leadership team that was renewed at least twenty times. When the remaining time was closer to the 400-year mark, the operations team started studying the handbooks, and eventually taught themselves how to fly the ship. Norman Link was on this team for several years and was elected by the team to be the ship's captain and pilot."

"Norm was the pilot who landed SC12 just 10 years ago. They had no communication and were not aware of other cruisers flying to Planet Alpha-1. A week after landing, they dispatched a small team to look around the area and locate some lakes with drinkable water. On the other side of a lake, they found a farmhouse. They spent several hours swapping stories. The farmer and family were offspring of the original families of SC15, that launched from SNEA two years after SC12."

"Families of the lost participants were understandably upset, and NSEA met all of the required costs, and offered our condolences. In addition, we have suffered from articles that blamed us for flying cruisers without a proper flight inspection and assigning pilots with insufficient training. NSEA decided to discontinue our program established for exploring Planet Alpha-1."

SC253. "Now, earlier this year, 410 years after SC12 launched, we decided to resume what we started so many years ago. We picked Larry Anderson, our best pilot, a very capable assistant pilot, a senior geologist, and 20 young geology graduates. Their first priority was to collect geologic samples and submit reports back to NSEA, They also had a second priority to do a search for crashed cruise ships and/or evidence of people who might have been in those ships.

"Well, on the first evening while circling the planet Alpha-1, Larry saw city lights. Larry continued a slow orbit and then landed the next morning, about seven miles away. Within the next hour, representatives of the personnel from SC12 and SC15 flew to Larry's spot by helicopter. Larry explained his mission, and they, in turn, told Larry their practically unbelievable stories."

Two days later, during a search for more people, they stopped to look in a large cave. They found painted pictures on the walls. There were empty paint cans on the floor, so this was definitely not painted by pre-historic beings. The next day they started a search and found evidence in the area north of the cave. Evidence of people fishing in the lake, and then a gravel road with automobile tracks. The tracks headed to the west, then disappeared under the trees of a forest. It was late in the day, and they decided to come back in the morning. Then, in Larry's buzz-craft, they drove the road on wheels, like a car, and found the Manager's office. Then we explained who we were and what we were doing, and David Parker told us his story.

CEO Continues. "I have given you a short story on each of the events starting 410 years ago. We could ask ourselves; Why did we not do a better pre-flight inspection? Why did we assign pilots without sufficient experience? Why did we not provide better comms systems, and why not provide back-up comms equipment?

We have made significant corrections to each of these. Our proof is we have had perfect results ever since. We can call that step #1.We just completed our step#2 with the very successful results accomplished by Larry Anderson and his team.

Larry's report on Priority #1 reported he had been in contact with personnel who are offspring from the two cruise ships. Also, he would continue a search for evidence of the third ship.

After Larry submitted his priority #1 report on their search for the ships and their survivors, or offspring of the survivors. Then he started on priority #2. He was provided with a summary of reports from Melvin showing the types of products available in several nearby areas. His senior geologist, John Collins. arranged their 20 geologists into 4 teams and flew them out to a large area for them to gather samples. John managed his teams for two days and covered 4 different large areas. He handed his report to Larry, who in turn, included these summaries in his report for his Priority #2 report. John commented that these are almost double the summaries he was familiar with back on Planet Earth.

Now, we are in the early stages of Step#3, helping to create a *geology business* on Alpha-1. NSEA has invested a good amount of time, money, and effort in helping to create a professional team, and we will continue helping on the business side.

Our President, Harold Phillips, reminded me he has a good head for setting up new organizations with new procedures Then he monitors performance and looks for ways to make them better. He is aware that the organization does not belong to NSEA. We helped them with the plans to get organized, provided ideas about extra training for the local geology workers. He also suggested they need to create a company and register the name. Something like Alpha-1 Geology Organization Inc.

Harold, or his assistant, will be the recipient of calls from clients, or possible clients, who wish to place an order. We need to know the products they are looking for, and the quantity. We NSEA is providing a space cruiser for one trip a month, at no cost. Harold will work that out with the clients, knowing that our delivery schedule is already set.

Melvin will keep track of which areas are 'best' for the requests from the client requests. His daily schedules will always give first priority to the client orders. Melvin has an assistant who ensures that the various products are dropped into the correct bins. Whenever we reach the completion of a client's order, an assistant will package the material and take that to Harold, who will enter packages in the space where he has a sign with the arrival date of the next space cruiser.

CEO. "Now, it is time to vote. The motion on the table is that Harold be released of his position as President of NSEA and he will take a position as 'Head Manager,' or something like that, with the Geology Group on Alpha-1. Any questions?" There were none, and he asked for a show of hands "if you approve." There were eleven hands. "That is almost all of you. So, the motion has passed. Would the two of you care to tell us why you disagree."

One was quick to comment, “I was concerned if we will be working without a President, or do you already have that worked out?”

CEO replied, “Yes, I will be CEO and President. Any more questions, or objections? None, do we have a motion to close this meeting?” CEO left the room, while the others stayed long enough to shake hands and say *I will miss you, good luck with your new job*, and *will you be coming back*?”

Harold said, “Here is what I told CEO, “If my first two months are looking good, I will move my wife to Alpha-1, and make that our ‘forever’ home. If those first two months do not work, CEO said I can have the job back.”

Harold flies to Alpha-1. He was packing his clothes, and other personal items, when he heard the President’s cruise ship landing on a pad close to his private door. He had called Harry the evening before to bring him up to date on his departure from NSEA. Then he asked Harry to be his pilot on what may be their last trip together. He picked up two suitcases, waited ten or fifteen minutes, then walked out about the time Harry was stepping from the ship. Harry chuckled and said, “Is that all you are taking? I expected more than that.”

Harold replied, “This is just the first load. I have more in the house. Would you come in with me for a few minutes?”

Harry's Smile had disappeared, "Of Course," and in they went.

Sitting there was a number of boxes, with his computer, printer, paper, a stack of books, and other stuff. Three trips. Harold hurried back to his wife for hugs and kisses, and then back to sit down next to Harry, his pilot and friend.

Chapter 11 – We're in Business

Flight to Alpha-1. Harold and Harry are best-of friends, and they had a good time talking, telling stories and jokes, and trying to play a card game, which didn't work out at all.

About one week into the 20-day flight, Harold sent a message to Norm, Larry, Melvin, John, Tom, Tommy, and Dave, including an explanation of what was agreed to at the NSEA Board meeting.

"I am hoping that you will agree to this. Please understand that my suggestions are not for my benefit. No, they are entirely in favor of the Alpha-1 Organization. I will be arriving in another two weeks, and I am asking that you find me an apartment. Nothing fancy, just enough to be my temporary home and office. We will need a large car, or small truck, or Larry's zoom-car to carry my belongings. Should we land by the hangar or at the pad on the other side of the airport? When the NSEA building is ready, I would like to have the office area on the bottom floor as my office and a place to keep records. Also, one of the upstairs bedrooms for a temporary apartment. Let's plan to do an assessment at the end of two months. If all is well, right after that, I will bring my wife to join me, and this will probably be our forever home. If things are not well, the CEO already said I can return to my previous position with NSEA. But the position will not be available to me after those two months.

Eight days later, Harold received a reply from Norm,

"Harold, we are so happy that you are joining us. As it turns out, the NSEA building will be completed several days before you arrive. So, you and Harry should land on the pad close to our new building. We can move your belongings on a pushcart. New subject. We had talked earlier about a celebration when the buildings are finished. Now we are thinking we should celebrate both the new buildings, as well as our new President. You are the man who will help us improve and strengthen our business knowledge, technical procedures, and the establishment of our new company, Alpha-1 Geology Organization Inc."

Arrival. It was a good flight. Harold had asked Harry to circle the area before getting too close. "I would like to have another look at the area on the northern side of the town. About one mile to the east, there is a factory where just last year they started making cars, and now they are preparing to build an airplane. Another mile east is a large cut into the mountains where then have been mining for iron for 300 years, and close by is the huge furnace for processing the raw iron into shapes the factory wants so it is usable by their people.

On the approach Harry checked in, “Air Control, this is Harry in our Cruiser, with the new President of Alpha-1 Geologists Incorporated.”

The response, “This is Air Control, hello Harry, you are cleared to land as you requested, and welcome to the new President.”

It looked like everyone from Our Town and West Village were there. Also, twenty people from the East had flown in on a zoom-car, thanks to Melvin the pilot. As Harold stepped out of the cruiser, we heard the loudest applause ever. Fifteen minutes later, the roar started to settle, and a group is carrying Harold’s belongings to the NSEA building. As he stepped onto the entrance steps, the applause restarted. We had noticed two empty flagpoles were standing nearby. Then, John and Dave appeared in front of their building and hoisted the flag. Harold then joined and the three of them saluted the flag. Another round of applause.

Several days later. Howard finally received his electronic copy of the magazine he had been waiting for. Howard’s buddy Jerry is head of the advertisement section, which includes advertisements for companies selling their geology products. Typical ads consisted of a company name, one or two products they would like to sell, with the related prices, and phone number so customers could call and place an order.

The ad prepared by Howard and Jerry follows.

> Announcing the opening of Alpha-1 Geology Products. We have been mining iron products for 300 years. Those products were all for our locals, but now our products will be available to you. We have iron products for sale now, and our price is listed below. As we start to accumulate other products, such as aluminum, silver, copper, and magnesium we will add those to our column. Oh yes, we are so far away our phones cannot connect, so we must communicate by e-mail. We look forward to hearing from you.

John Collins had focused his business on those streams that are close to home. The average days were about 5oo pounds of rocks, and nearly half of that was iron. Aluminum and magnesium usually scored about 20 pounds. Copper was inconsistent, showing up 4 times with about 10 pounds total. They saw traces of aluminum occasionally, and. only a few ounces. They never saw magnesium.

John called Melvin to see if he would recommend, “that place where you went down inside a one-time active volcano.”

Melvin said, “That is an excellent idea. But don’t go there without me. I will ask Larry if I can borrow his buzz-craft so I can buzz over there. Then I’ll pick you up and I’ll show you around. How about tomorrow morning?”

The next morning, about nine, Melvin phoned John, “Good morning, we just left, so I should be there in about half an hour.”

John replied, "I'll be ready. Who is that coming with you?"

"That's Harold, our new President. He is begging to see how we do things. That helps him understand what we do."

John was standing outside of the NSEA building, waving when Melvin made the landing. John climbed in and they chatted for a minute. Melvin pointed to the open seat in front, John started to sit, then said, "Shouldn't our President be in the front?"

Harold said, "You two are working and need to be up front. I am just a newbie, and observing what you do,"

Melvin said, "We start out heading due-West, and it is about 100 miles away. We are at 2000 feet, so you can watch the horizon ahead of us. There, now I see the outline of mountains, maybe 40 miles away. Back at least a hundred years ago, a mountain blows its top and becomes a volcano. Some of these keep going and bring up more that plugs the hole. These mountains usually have a pointed top. But others quiet down and just leave it as is, and usually end up with a flat top. Larry was the pilot and we started looking at the flat tops. He asked me where I wanted to go. I saw one of these had a lake in the middle – interesting but no help. Another had trees in the middle, and that might mean there was an island. I pointed and said *that one*. Well, that's the one we're going to now."

Melvin landed, moved to the same place where he was about 5 or 6 months earlier.

"You can see, there is no lake. The 'hole' goes down about 40 feet, and the trees are growing from there. I wanted to go down. Larry said he had a heavy rope in the back. He opened the back of the buzz-cart and pulled out a large coil of heavy rope, and a harness for me to put on. Then he hooked the rope to the harness. Larry's a pretty smart guy. He was concerned that if I went over the edge the rope would be cutting out the dirt and perhaps it could loosen a rock that might hit me on the head. The other end of the rope is attached to an electric winch. He set it on the ground, then gathered some large rocks and several heavy limbs to keep it from sliding around. Larry sat down in the buzz-craft and touched the climb button and started pulling me slowly up to five or ten feet above the ground. Then he moved over so I was about five feet inside the outer edge of the hole. We both had phones on so I could say something if he needed to move a bit. Larry had a device in his hand that controls the winch for my up and down trips. We also had a third party, sitting in the buzz-craft, with responsibility to keep an eye on me as I was going down, mainly to watch for anything that I might not be able to see. That is partly why I brought Harold."

Harold commented, :I was afraid there would be a catch to this. Just kidding. I am so glad you brought me along. Honest."

Half an hour later, John was hoisted up, and he was really excited, "There is a wonderful collection of metal sources. But getting to it would be a real chore. We would need to have several winches for the many lifts of 100 to 200 pounds of rocks, or slices, or chunks. Then we would have a series of flights to carry all this back to our headquarters. And there we have to sort everything according to the appropriate nomenclature."

Harold jumped up, "And then you will need to send me email with your daily results, so I can update our monthly advertisements."

Melvin commented, "But you have overlooked the most important part of this. You might start digging holes into the wall so you can reach more products. And then one day the side you are working on collapses, and maybe buries some of your workers. An alternative might be to start collapsing one of the sides from the top. Start out by removing some dirt from the top, then scooting all that away from the hole. After you are down a few feet, then you provide a stick of low power dynamite. Then *bang*, the side is gone, and we have a large pile of dirt at the bottom. Then you push away the dirt to the other side, and you probably have access to a large amount of iron. There are other possibilities. How about this one. Take out the trees, then place some dynamite around the top and collapse the entire wall. Most of the dirt will collect at the center, and you have a lot of iron and other products. Just think of the iron pit that the folks from Our Town found 300 years ago, and it is still alive. Here is something to think about. Should you invest in some machinery to help with those big pieces?"

Harold commented, "Here is something I found out about just a few days ago. I was looking at our zoom-car and thinking that would make a good truck. Take out the seats and that could work. I contacted one of the engineers at their headquarters. He and I had several meetings when we were talking about buying our first zoom-car. I asked him straight out what it might cost if we wanted to buy 2 zoom-trucks. He laughed and said, *"Not Necessary"*. When they were building the first zoom-car, they decided to make it easy to convert to a truck. Look at the legs of the seats and you will see the gadget that holds the seats down. Just turn the lever and you can pick up the seats and move them elsewhere. You can stack them in a corner or remove them. The engineer apologized for not mentioning this to me earlier.

John said, "That works for me. I'll ask Dave to deliver our team to their site tomorrow morning. Dave is a pretty good pilot."

A few days later, Harold was writing his advertisement for the next magazine issue. He thought the rest of might want to have a look.

My draft for the next issue, "Our 300-year-old iron pit is doing great. We pulled up a thousand pounds this month, and now we can sell up to 1500 pounds. In addition, we have started investigating some other possibilities. We will let you know next month how that worked out. We have not had any luck with the other metals. 5 pounds at a time, here and there. But we will keep our eyes open. More next month."

Then Harold sent an email message, "To the senior folks of Alpha-1 Geology Products. I suggest we have Board Members. For the West area: Norm and Melvin. For East we have John and Dave. And then we have Larry and me. I suggest we have Board Meetings once every week. How about Wednesday at seven pm? And how about this meeting at the East NSEA building, in the office on the first floor. Larry, will you be able to fly us westers to the East building. You have the buzz-Craft and that is only a 30-minute trip. Joan and Dave would be 4 hours each way, in their zoom-car. Please let me know if you can or cannot make the meeting. Also think about my suggestions. You can let me know now or wait for the meeting."

Wednesday. Four of us are in Larry's buzz-craft, heading for the East NSEA building, a 30-minute flight for this 4,000-mile flight, just chatting away.

Melvin said, "I have wanted for a long time to search those mountains behind Our Town. But then I set it aside for something else. And then I forgot about it. I will set that up as number one priority for tomorrow morning."

Harold commented, "Hey, you will have to repeat this for our meeting. But you will need some details. Do you have any others for tonight?"

"Well yes, but that one is for east. You remember where Larry stopped so I could look for paintings in the cave. I believe we already did a walk in the stream for several miles. That needs to be repeated every two or three months. The other one is the cliff. "That's a good one. I'm surprised John hasn't picked it up himself.'

Melvin commented, "Actually I was going to wait until he finishes his report, and then I can bring it up. I will do it as a "*suggestion*," so that way I am not taking attention from his talk."

The meeting went just fine. Everyone agreed to everything that Harold had suggested in his message 3 days ago.

Melvin was ready to talk about his West team, "During this past month, we have focused on training our beginner geologists. They have found a ton of rocks, but nothing that will bring a nice price. Starting tomorrow. we will be searching the area just east of Our Town's iron-processing plant. I noticed that the hillside there looks like the scenery around the 300-year-old iron pit that they are very proud of. If we don't find anything interesting, we will move another twenty feet to the east. "Melvin gave a big smile and sat down.

John was looking forward to this. "Well, we did nothing that will sound impressive, other than a few hundred rocks. Melvin was very kind to show me a possible that he saw several months ago. Larry and Melvin were there looking at a mountain area near the east area. This was an inactive volcano. It has a 40-foot hole in the middle, and trees growing up from the bottom. Larry had the equipment to lower Melvin to the bottom. Melvin was impressed by all the pretty rocks and big chunks of iron and some sheet iron which is not found often. After nearly an hour, he said he was ready to go back up. Once up, he said, "It's spooky down there." Like I said, it was impressive, but I am not in favor of searching more down there. However, I do want to investigate some of the other mountains in the area."

John continued, "Our first endeavor will be closer to the East headquarters. Melvin also has credit for finding this. We were in the buzz-craft and came across a large cave. Melvin talked Larry into stopping and Melvin went inside. He was thinking about the painted pictures back on planet Earth that were painted by pre-historic cave people. He was inside for only a few minutes, and he hollered to us, "I found a picture painted on the wall. We will need to notify the experts so they can study this. Then, as he was coming out of the cave, he handed his phone to Larry. Almost immediately, Larry laughed and handed the phone to Norm. He also laughed, then, "Didn't you see the empty paint cans on the ground? Those were definitely not pre-historic people." Melvin was embarrassed and said he was sorry for taking our time on this. Larry was quick to point out, "Melvin, this is very important. We were looking for evidence of people in the area, and you found it. We thank you". The next day, Larry brought us back to the area to find more evidence of people. We had divided into four search teams, and mine just happened to go along a stream that flowed down from some large hills several miles away. The stream was loaded with very colorful rocks. The shape of the land was unusual, with the land on our right side nearly a hundred feet higher than the land on our left. The high side had a very long cliff that dropped down to the stream. We traveled the stream for several miles, where we noticed the water was getting deeper, and rough. The rocks were larger, and we saw large chunks of iron in the cliff. Well, that stream, and the cliff, are worth a closer look. That area is where my team will be tomorrow morning."

Harold could not have been happier than this. He thanked Melvin and John for their presentations. He also thanked Larry, "Larry, we could not have done this without you. I hope you will be with us forever."

"Before we quit, I want to say something about what I do. Well, as you know, NSEA is a government organization, and we don't make decisions to earn more money, but. I have a knack for understanding those things. In my early years with NSEA I took some courses in geology, and I happened to find that magazine that has a page or two about those companions that are doing well. I always wanted to jump in there, but of course we can't. That was a large influence to support my decision to leave my position with NSEA. It is my wish to do those things that will succeed in turning our Alpha-1 Organization into a very successful business. Any questions? No questions, so we are dismissed. Our next meeting will be a week from today. Same time and same place. I hope Melvin and John will tell us more about how their businesses are doing."

Two Months. The NSEA Board of Directors held a meeting to determine if Harold has performed in a manner appropriate to remain there permanently. The vote passed and he remains in that position until he resigns.

The Board also voted to promote Larry Anderson to President of NSEA.

The Board also voted to position Harry Treston, known earlier as the President's Pilot, to become a resident of Alpha-1, and function as Harold's pilot. Harry has already agreed to this.

Harold received the information 4 days after it was sent, and he immediately forwarded this to all of our citizens in Alpha-1.

He also phoned Larry, "Congratulations on your well-earned position. Are you going to head back pretty soon to Planet Earth?"

Larry replied, "Well yes. I am finished here, and I'm certainly anxious to report, *'Ready for duty'* to the CEO."

Harold continued, "I am moving my wife to Alpha-1 and can use a ride to Planet Earth so I can help Sally, my wife, pack things we are moving. Hey, I guess you will be moving into the President's house, right?"

"Well, I never thought about that. But wait, is it possible the CEO will take that. Do you know?"

"Well, I don't know for sure, but he currently lives in a large mansion on the other side of town. He comes from a very wealthy family, and I'd bet he avoided the President position because he didn't want to move. Now don't you ever mention I told you that, OK?"

"OK, I won't tell. I was planning on a brief get - together with my people at ten in the morning. We will be at the picnic tables, not more than an hour, then it's just a short walk to my cruiser.

"That's fine with me. I'll be there at ten, and then we have 20 days together, we can have fun playing cribbage."

"That sounds good, but you'll have to teach me how."

Harold sent a message to Harry. He told Harry he was delighted that he agreed to the CEO's change. Harold gave him the details for helping to move his wife's belongings to his cruiser, and then fly to Alpha-1. Then, after his cruiser is parked, Harry would help move her belongings into Harold's bedroom. Then Harold told Harry he had already planned for Harry to have the small apartment next door to Harold and Sally.

The next morning, the West NSEA picnic tables were buried beneath those who showed up to say good-by to Larry, now leaving to step into his new job as President of NSEA, on Planet Earth. They watched their lift-off with heavy hearts and sad faces because they may never see him again.

Twenty days later, Larry landed on the pad closest to the NSEA President's house, his new home for a long time. Harry waved as Larry and Harold stepped out of the cruiser. Sally had watched the landing from that place on the roof that Harold prepared several years earlier for watching landings, and she just now realized this is the last landing she will ever see. She was happy, but she was sad too,

Sally went down to their living area, and the others, Harold, Larry, and Harry came in at about the same time. The furnishings came with the house, so they only had to take down her personal belongings. This was a fairly easy task. Harry's cruiser was standing on another pad, only a hundred yards away.

Two hours later, everything had been placed where Harry wanted them, and they all waved goodbye to Larry. Another happy and sad situation. Harry had the engines started.

After another four hours, Harry said, "Pretty soon I will need to start our acceleration. This is called 3 jumps. We do this with all of us sitting in a regular seat, and strapped in. I will give you a one-minute warning, and then 3 –2 –1 go! And you will feel more acceleration than you have ever imagined. This will plant you in your seat, unable to move your arms or legs. The first jump takes eight minutes. That is a long time. Then we pause for five minutes, and the second and third jumps are not as long. When finished, we will be flying thirty times the speed of light. Why do we do this? In the earlier days, the trip to Alpha-1 took eight months. Now, with 3 jumps, we do it in twenty days. Here is the bad news. We have to do it again so we can decelerate. It is pretty much the same process, except I have to flip the ship around, so then we are cruising backwards. We will do this the day before landing. Any questions? Take your seats and hang on."

Except for the two jump events, it was an interesting trip. Harold enjoyed describing what has been happening on Alpha-1. Occasionally they played cribbage or took a nap. Harold had been a cruiser pilot for a couple of years, then he started looking for management opportunities. Managing was, and is, the thing for him. Harold's pilot license expired a long time ago, but he certainly knows how to step in for an hour or two when a pilot needs relief. It ended up that Sally was good at keeping an eye on whoever is in the pilot seat and several times she had to say, "The pilot needs a relief," and no one argued.

It is now landing day. Harold had told Harry where the landing pads are, and he should use the one on the West side. That is the closest.

Harry called AC, "Hello Air Control, this is Harry Treston, in a space cruiser, approaching from the west. I have orders to land on the pad on the west side of the NSEA building."

The AC response, "Hello Harry, is the new President with you?" Harry replied, "Yes, he is, and so is his wife, Sally. Am I cleared?" She came back, "Yes, you are cleared to land. And welcome Harold and Sally, I look forward to meeting all of you."

Harold said, “Nice landing. It is a little after three and the place is quiet. Maybe we could have a ‘welcome’ about 6 o’clock. I can arrange for drinks and snacks on the picnic benches.”

Harry and Sally agreed. Then Harold opened his computer and sent the message to his key personnel, Melvin, John, Maynard, Tommy, Tom, Dave, and David.

Harold found two of the duty push carts, and they started loading them. They pushed the carts to the building, around the building to the entrance, then carried everything up the three steps, and unloaded inside the entrance door. Only two trips were required. Then they carried everything up to the two temporary apartments on the upper level.

At six o’clock, Harold, Harry, and Sally appeared briefly on the roof level. They waved, and then went down and appeared at the entrance door. There was good applause. After a couple of minutes, Harold waved that was enough. He held up an arm, signaling for quiet. After all was quiet, Harold smiled and said to the crowd, “Thank you for the welcome. I left here forty days ago with the purpose of bringing my wife to live here. with me. Forever.”

Harold turned to her, “And now I am introducing my wife, Sally, my favorite person.” After the applause settled down, Harold continued, “This other person is Harry Treston, known in our circles as The

President's Pilot. He has been with us for a long while. He has been with us so long; we have become more than just friends. Now he lives with us, close as a brother. And now Harry also has the responsibility of transporting our geology products to Planet Earth when we make a sale. By the way, Harry and I will be sharing the office on the first floor of this NSEA building."

"We thank everyone for showing up today. We have drinks and snacks waiting for you on the picnic tables. Help yourself. And enjoy yourselves.""

Chapter 12 – Moving In

It was just a few yards from the picnic tables to the NSEA building.

Alone at last, Harold showed them around the first floor, "This large open area is a place for people waiting for a flight to come in, or a group meeting, or a celebration. This is a good place for people to welcome guests. This next room will be our office. Harry, this is for you and me. I would like to keep all of our papers here. We will have two desks, and a file cabinet for each of us, so we can have files on everything. This other empty room is for a collection of tools, a couple of hammers, several screwdrivers, wrenches, and boxes of nails, nuts, and bolts. This is intended to help someone who may have to do a quick fix on something. We can also keep the geology products here, ready to send to a customer on Planet Earth. I had earlier been thinking we could schedule trips for one each month, but that does not work. Each trip takes 20 days, each way. So, we will change that to a trip every other month. That is, if we have a lot of business. Enough of that. Let's go upstairs."

Harold took them up the stairs to the second floor and their temporary apartments. They were both the same. Each had a good-size room with a bed, a lazy chair, several sitting chairs, and a small table. The next room was a combined kitchen and dining room. The other room was just large enough for the commode, shower, and sink.

Sally was not impressed. She asked, “How long do we have to be here?”

“I don’t know yet. I have to call the NSEA carpenter. First let’s go up to the roof. We have a good view from there.”

“Now. Look over there, to the South, that is the airport, almost a mile away. The large building there is the hangar. All of our buzz-crafts and zoom-cars are kept there. There is also a helicopter. Plus, there will be an airplane in another month or two. Now, look to the West. See that nice-looking house? That is our new home. Only a quarter mile away from here. I was about to say we should walk over there. But let’s wait a few minutes and I will give him a call.”

Harold was on his phone for fifteen minutes.

Finished, he smiled and said, “The carpenter needs two more days, and then he is finished. I asked if I could have a walk-through, and he said no, the other NSEA building is behind schedule, and he needs to hurry over there as soon as he can. I will try to arrange for us to have the buzz-craft that morning. That way we can ‘buzz’ over there and have our own walk-through, then back here to pick up our belongings and deliver, My guess is that will probably take two trips.

Another Moving Day. Harold, Sally, and Harry had packed their personal stuff the night before. They were only moving two hundred yards, but Harold had figured this would still be a large effort for them to move everything into their new home. Harold asked Melvin if they could 'borrow' a couple of the younger guys to give us a hand. Well, Melvin and about ten of his workers showed up. One group of Melvin's people carried everything down and laid them down at the entrance door. Then the next group loaded them into the push carts, pushed them to the new home, and then group three carried armloads into the new home. Group one was waiting and then carried the pieces to the correct rooms. Then Harold, Sally, and Harry were standing by to

say, "That goes there," for each load.

When finished, Melvin pulled his group together in front of the front door and told Harold his guys wanted to say good-bye. Harold, Sally, and Harry quickly appeared, and Melvin's group said, in unison, "Congratulations on your new home."

And then, Harold, Sally, and Harry said, in unison, "We thank you for your help."

Melvin huddled his group for a short minute, and then, "Best wishes to you in your new home." Then they waved, and Melvin took his group to his zoom-car. And they were gone.

Harold needed some assistance for establishing his communications, in both his house and his office in the NSEA building. He didn't know who to contact, so he called Norm for help.

Norm explained, "We don't have house phones anymore. We use cellphones. and they are all wireless, so you don't have to be plugged into anything. Except, of course, you need to plug it into power now and then, to charge up your battery. Television is also wireless. We only have one station, so that keeps things nice and simple. Our major connection is email. Again, this is wireless. If you send an email to someone back on Planet Earth, it still takes more than two weeks for delivery. But, for us with *NSEA* included in our addresses, it is only one or two days."

West Geologists. Our geologists have been doing a great job. Harold will be updating our ad for the geology magazine in another week, and the numbers are very good.

Melvin is currently focused on his new iron area, which starts 100 yards to the east of the 300-year-old mine they call *The Iron Pit.* Their workers are all employees of the Our Town Factory. They are professionals, and they work three 8-hour shifts, around the clock, six days a week. Asked why they work so hard, one of the workers said, "The factory is gearing up to start producing one car a month, and they just increased our pay to $25 an hour, and the shift-bosses get $30."

Melvin was anxious to talk to his team. He walked his team 100 yards in the eastern direction, stopped, and then said, "Well gang, the Iron Pit has been very successful for a long time. By comparison, we are just beginners, but I want you to work hard and reach that professional level just as soon as we can. I am putting up a mark here for team 1, then I move over 10 feet and mark this for team 2, and another 10 feet, mark for team 3.

His workers got started right away, shoveling dirt and other small stuff out of the way, and picking away at the heavier chunks to break them into pieces that are carriable. Marvin made a note to pick up some more tools, especially 3 heavy-duty sledges, and 3 heavy duty carts.

While watching his workers, he realized he was going to have several more jobs. As the workers remove dirt and unusable rocks and create piles on the road, Melvin just visualized converting his zoom-car to a truck, and then carrying the junk to another area. And he thought, "Where can I dump the junk? Do we have any designated land-fill areas? Also, at the end of each workday I will have to pick up all the salable materials and take them to wherever Harold wants

East Geologists. John parked his zoom-craft alongside the stream. Then he split his group into two teams.

"The first team will be walking in the water, heading upstream, while looking for rocks that appear to have metal. You can pick up everything that even looks like it might have signs of metal. Before we leave, we will give them a closer look and sort them into three groups: one for iron, one for other metal, and then we throw the others back into the water. Walking against the current is work, so you should take a break every hour. We will have a 30-minute lunch break at noon, and at three we can start sorting our rocks. "

"The second team will be looking for larger clunks of iron. We will be on the high side, and use ropes to scoot part way down the cliff, looking for traces of iron. I have some heavy duty spikes we can pound into the ground. And I have several strong ropes that will help you go up and down. If you find something interesting, you might have hit pay-dirt. Stay where you are, and we will talk about what to do next. If you find nothing, we will pull you up and move you to another spot a few yards up the line. Remember, the ancestors of SC15 opened their very successful *iron pit* 300 years ago. Melvin and his people are looking for iron in that general area, Are we up to competing with those guys? Let's go to work.

It is a busy month. On the first day, Melvin's team came across four chunks of iron. Melvin made their delivery to Harold, and their results measured 17 pounds. That was nothing to be proud of, but it was something. This was their beginning, and they were happy and proud.

The next day started out pretty much like the first. And then, maybe an hour later, one of the workers yelled out, "I've got something here! I can use some help." They could see this side of another iron chunk, but this was larger than the ones yesterday. Four guys are working on this. One on each side and one on top, all with their rock hammers. A fourth worker was working below the rock with a shovel and a crowbar, trying to loosen things away. The guy on top grabbed another crowbar and started prying from the top. Then they heard a creak and a crack, and they all pushed away just in time. The giant rock had moved out several feet. We could see behind the rock some of the dirt and other small stuff was still moving downward, filling the gap behind the giant rock.

Melvin was nearby and came over to see what was going on. "This looks like a 'good one'. I recommend we take a break. Maybe half an hour, to let everything settle. Then listen for any more warnings."

By mid-afternoon, they had pulled three more large rocks and a dozen small rocks. They had no way to estimate the weight of the rocks, so Melvin asked one of the workers at Iron Pit to give us some estimates,

He spoke. “You guys did pretty good, for your first day. That big one is about 250 pounds, and the other three range between 100 and 150 pounds each. The little ones are about 10 to 20 pounds. The four big ones are definitely iron. The little ones all appear to be metal, but you will need an expert to designate the type. “

Melvin picked six of his men to help load these into the zoom-car. Then he re-installed six seats, and they zoomed to their ‘barn’ to ask their expert for the specific designations, and where they should put these.

John, at East, was also having a busy day. The stream was loaded with 10 to 20 - pound rocks, including all of the metal designations, even two, 5-pound gold rocks. The side of the cliff was exciting. The first one on the cliff pulled out six 20 to 40-pound rocks. Two other workers performed about the same, and another hit the jackpot, with a 200-pound iron rock. He called for help and John set up another spike and rope for another worker. The two of them inserted some spikes and fastened several ropes to hold it where was, Then John dropped down the lower end of two more ropes. The workers below fastened all the ropes and John picked 12 more workers to become pullers at the upper ends of the ropes. After ten minutes of grunting and pulling, they had their prize iron rock. John removed the unnecessary seats to make room for their products. John and workers loaded their products for today, then John headed for the barn.

Today is Friday, and we provided the numbers to Harold so he could prepare his new ad in the geology magazine on Planet Earth. Our next step was to relocate their barn from an isolated location outside of town and position it close to the NSEA building. Harold released the geology *expert* who was hired temporarily to identify the various metals we might be working with. Then he rented a large truck for Melvin's men to use in disassembling the barn and then re-building the barn at a position close to the NSEA building and a launch pad. The next event in their schedule was for Harry to pick up the sold products in his space cruiser and deliver them to NSEA Headquarters for passing on to the customers on Planet Earth.

The End

Key Players in this Story – by order of introduction

CHAPTER 1	
Capt. Nick Logan	Pilot and Captain of SC12
Cdr. Bert Castner	Operations Officer
Charlie Carter	Medic (former military medic)
Jack Brown	Doctor (MD)
Hank Hodge, Mr.	Space Engineering consultant (theoretical)
Betty Hodge, Mrs.	Space Engineering consultant (practical)
CHAPTER 5	
Capt. Norman Link	Pilot and Captain of SC12, 400 years later
Cdr. Bob Calhoun	Operations Officer
Tom Kennedy	President of the Board
Lt. Jerry Wyatt	Director of engineering
Melvin Sherwood	Senior geologist
Clyde Holdren	Senior scientist
Lt. Patricia Andrews	Manager of Computers
CHAPTER 6	
Brad Wollam	Farmer
Mr. Mayor	Mayor of Our Town (Maynard Jones)
David Andrews	Helicopter Pilot
CHAPTER 7	
Larry Anderson	Pilot and Captain, SC253
Jim Womack	Operations Officer, SC253
John Collins	Senior geologist, SC253
Maynard Jones	Governor, West area
Tommy Smith	Mayor, Our Town
Tom Kennedy	Mayor, West Village
Dave Parker	East City Manager
CHAPTER 9	
Harold Phillips	President of NSEA
Sally Phillips	President's wife
Harry Treston	President's pilot
CHAPTER 11	
Harold Phillips	President of Alpha-1 Geology Inc
Harry Treston	President's pilot, in Alpha-1
Larry Anderson	President of NSEA

Definitions

Term	Meaning
BB	Basketball
3-D Basketball	Basketball, played in zero-G. We can "float."
Alpha-1	Our destination, a planet, outside our solar system, 4 light-years away
AVC	Combined audio/visual/computer equipment – in all bunkers
Bunker	Quarters for 1 – 3 personnel on SC12
Buzz-Craft	Nuclear-powered. Seats 4, plus 6 more in side-car. Top speed 8,000 mph
Zoom-Car	Similar to Buzz-Craft, 20 passengers, Top speed 800 mph
E&M (Department)	Engineering and Mechanical Department (Director of engineering)
Head	Restroom
HQ	NSEA Headquarters, on Mother Earth
L	L1 equals the speed of light. L2 equals twice the speed of light. Etc.
NSEA	National Space Exploration Administration
One G	One G is the same as the normal gravity on Planet Earth
POD	Combination chair, lounge and bed – in all bunkers
Pri-comm	Speaker system, heard throughout the ship and in the AVCs
Zero G	No feeling of gravity, we can "float" through the air
SC12	Our ship, Space Cruiser 12. 250 on board
SC15	2nd ship launched 2 years after SC12. 250 personnel
SC16	3rd ship launched 1 year after SC15, for search and rescue ' 40 personnel
SC253	The recent Space Cruiser, launched 410 years after SC12 – 120 personnel

Made in United States
North Haven, CT
04 November 2026

21766916R00303